The Nobel Banquets

A Century of Culinary History
(1901–2001)

I0809212

The Nobel Banquets

A Century of Culinary History (1901–2001)

Translated from the Swedish by Michael Knight

Ulrica Söderlind
University of Stockholm

World Scientific
NEW JERSEY • LONDON • SINGAPORE • BEIJING • SHANGHAI • HONG KONG • TAIPEI • CHENNAI

Published by

World Scientific Publishing Co. Pte. Ltd.
5 Toh Tuck Link, Singapore 596224
USA office: 27 Warren Street, Suite 401-402, Hackensack, NJ 07601
UK office: 57 Shelton Street, Covent Garden, London WC2H 9HE

This English edition is a translation of the Swedish edition "Nobels middagar — Banketter, festligheter och pristagare under 100 år" published in 2005 by Carlsson Bokförlag.

Copyright © 2005 by Ulrica Söderlind

English language translation copyright © 2010 by World Scientific Publishing Co. Pte. Ltd.

All rights reserved. No part of this book may be reproduced in any form or by any electronic or mechanical means, including information storage and retrieval systems now known or to be invented, without written permission from the publisher.

A catalogue record for this book is available from the British Library.

ISBN-13 978-981-4313-11-7
ISBN-13 978-981-4317-97-9 (pbk)

Printed in Singapore by Mainland Press Pte Ltd.

To Wilhelm Odelberg

In memoriam
1918–2002

Contents

Foreword

When I was invited to be Restaurateur at the City Hall Restaurant in 1989, the Nobel Banquet was the great challenge but also a great cause of alarm. Would I be able to manage this huge responsibility or should I decline the offer because of the Nobel Banquet? How is it possible to cope with the 330 staff or more in this gigantic show? How do you keep the food hot, keep to the strict serving timetable, manage to work together with all the other people involved, meet all the wishes for special diets, provide drinks for about 1,350 guests and organise all the porcelain and cutlery, which has to be shining clean? Will I cope with the stress and publicity and how will my family cope with having me shut away in my own little world for months before the event? There were many questions to think about, but I quickly realised that it is thanks to all the fine, competent staff and other organisers that everything works perfectly. I am both glad and proud that I accepted the post and that I had the honour to organise the Nobel Banquets until 2002.

During all these years, with the climax in 2001 of the centenary celebrations, I have faced many problems and had many thoughts about my role as restaurateur for this famous banquet.

Is there any banquet, national or international, that is more well-known than the Nobel Banquet? Have the Nobel Banquet menus influenced other similar arrangements over the past 100 years, and if so, how? What is served at the Royal Palace on December 11 when their Majesties the King and Queen of Sweden invite the laureates and other honorary guests to dinner?

These reflections of mine have now found answers, thanks to all the time and effort that Ulrica Söderlind has devoted to collecting information, not only about 100 years of Nobel Banquets but also about the dinners that the King and Queen give at the Royal Palace on December 11. Ulrica's never-ending research in the archives of both palaces and more humble abodes has benefited us all. We now have fresh knowledge about the first 100 years of the Nobel Banquets.

Ulrica, whom I first met when she started to work as the cold buffet manageress at the City Hall Restaurant in 1990, has fulfilled a dream that

Alfred Nobel photographed the year he died, 1896. © The Nobel Foundation

I and many others have had for a very long time: to present all the menus from 1901 that were missing from the archives, with facts and comments, to read what was written about the Banquets from the beginning of the 20th century onwards and to see how, year by year, what began as a smallish dinner at the Grand Hotel for 118 gentlemen, who paid 15 kronor each, has grown into the world-famous Nobel Banquet. Fantastic!

Producing the menu for a Nobel Banquet is a long process. The contents should have a Scandinavian character in which, as far as possible, Swedish ingredients form the basis of the menu.

An important factor that has to be taken into account is that the kitchen is a long way away from the Blue Hall, where the food is served — in fact, it is four floors above. We are grateful today to Ragnar Östberg, the architect who designed the City Hall, that he included lifts between the kitchen and the Golden Hall, but to let a lift go down to the Blue Hall would have helped us greatly in our work. Those of you who are familiar with the City Hall will know that there is, in fact, a lift down to it, but Östberg had the floor in the Blue Hall's south cloakroom made of cobblestones. Any of

Lars-Göran Andersson and Ulrica Söderlind conversing in the Golden Hall, Stockholm City Hall. © Torbjörn Boström

you who have tried to push or pull trolleys loaded with glass and porcelain across a cobbled passage will know that it is no easy task; in all probability most of the glass and porcelain will get broken. Another not insignificant factor is that the distance between the Golden Hall's serving counter and the Blue Hall, via the Blue Hall's great staircase built of Kolmård marble or the staircases in the north and south stairwells is a considerable way for the waiters and waitresses to have to walk, carrying in heavy, shining silver serving dishes full of food or carrying out dirty porcelain and cutlery. That's a lot of walking before the day's work is done.

In addition, the menus have to be designed so that about 1,350 people can be served in as short a time as possible, without any unnecessary and time-consuming serving problems. After months of pondering and a number of different proposals presented to the Nobel Foundation, it is time to decide on the contents of the menu at a trial dinner held early in the autumn. Twelve people, sometimes a few more, representing the Nobel Foundation, the City Hall Restaurant, the Catering Academy and others involved in designing the menu sit down to eat, enjoy, discuss and comment on the dishes before finalising the menu and returning to their daily chores. Remember, the menu is kept a secret until 4 p.m. on Nobel Day.

You are now holding in your hand the results of Ulrica's work, and her efforts give us a splendid insight into the first 100 Nobel menus and the menus served on December 11 at the Royal Palace.

I, and so many others who have been involved with the Nobel Banquets, are deeply indebted to Ulrica for her work. It is extremely gratifying that the first 100 years of the Nobel Banquets have now appeared in print, and we look forward to yet another 100 years of Nobel festivities, with the Banquets at the centre of attention.

Lars-Göran Andersson
Stockholm, Sweden
December 2004

Preface

There are probably as many reasons for writing a book or books as there are writers. The reason for my writing this book about the first hundred years of the Nobel Banquets came to me when my daughter and I were watching the prize-giving ceremony in 2001, not so much because it was the centenary as because my daughter, then ten years old, said something that stuck in my mind.

When the prizes were being handed out and the television commentator remarked that one of the laureates did research on the human brain, she said: "Good, keep doing that brain research, it's needed." This comment came straight from the heart because a close member of the family had suffered a stroke and had to spend time convalescing, which had repercussions on the rest of the family. That was the very moment when I got the idea to write about the Nobel festivities, not only because it was the centenary year but also because no one had previously written about the way the banquets had developed. I described my plan to my former restaurateur, Lars-Göran Andersson, who thought it was a good idea. With his help, I then got in touch with the Nobel Foundation and they, too, liked the idea. After that I only had to start writing. Another reason for choosing to study the banquets was my own personal involvement as I had worked in the kitchen there for a number of years, and food and drink have been my great interest ever since I was about ten years old.

This book is dedicated to the memory of Wilhelm Odelberg, that wonderful man who, to my very great surprise, also took an interest in my work. If ever there was a man who deserved to be called "Mr Nobel", it was him; he attended all the banquets between 1959 and 2001. Wilhelm was a fantastic source of information, particularly for Chapter 4, which describes the development of the banquets, since he had been present at so many of them. When he fell ill, he telephoned to tell me and we agreed that he would continue to help me, but as he could no longer manage personal meetings, we developed a system by which I wrote down my question(s) on a postcard and sent it off and he telephoned me when he felt strong enough — and that was usually the same day he received the card. What turned out to be our last conversation took place on the day before he died, when he was still in good spirits and was himself working on an article. Besides his knowledge of the banquets, he was a very humble man who kept himself up to date on current affairs in the world and who had a broad and open mind as well as great human warmth. Since we had both studied history and archaeology for our degrees, our conversations were always about much more than just the banquets.

The book is organised as follows: After this introductory preface, there is quite logically a chapter that describes the background to the Nobel Prizes. The next chapter is called "The Gastronomic Man" as it deals with the causes and factors that influence people's choices of food and drink. This is followed by Chapter 3, which discusses the gastronomic mix of countries that were present at one and the same time in the Blue Hall of the Stockholm City Hall at the 90-year and 100-year jubilees. Chapter 4 presents the development of the banquets themselves over the first 100 years, while Chapter 5 deals with the menus. This chapter also includes the Court's banquets for the laureates on 11th December because the first of them was given in 1904 and they have been given ever since and are closely connected with the Nobel festivities. Another reason for including them is that I wanted to compare the menus for these two banquets since both are given in honour of the laureates. I have divided Chapters 4 and 5 into ten-year periods for the simple reason that it was easiest to present the material in this way to make it more readable and more easily comprehensible for the reader. I also chose this system to make it easier for readers to find any particular year they are interested in. The menus for each ten-year period, however, are presented together, but I refer anyone who is interested in one particular year to Appendix 1, where all the menus are listed year by year in French. Chapter 6 is a short conclusion. It is also quite possible to read each chapter separately, if you so wish.

No book writes itself and even though only my name is on the cover, this book would never have materialised without the help of all the people who answered my inquisitive questions. Writing a book is not entirely painless and is sometimes a complicated process; at times one finds oneself lost in an unbelievably muddled world of ideas that manifest themselves in various ways.

First and foremost, I want to thank the people who are close to me every day and who are still there: Iván and our daughter Rubi. Many thanks for putting up with my whims and fancies and all that they have involved. My thanks to my sister Helene Söderlind, who is geographically quite a long way away, for patiently reading the first drafts that I sent her. It couldn't have been easy for her to read these chapters since her personal interests have never been in the kitchen region — and never will be! I thank Johan Söderlind for his help with computers and their technicalities, which I do not master. My grateful thanks also to Marianne and Ove Söderlind (my parents) for their never-failing interest in all that I have done.

The embassies of the following countries are sincerely thanked for having provided me with invaluable information concerning their countries' eating and drinking customs: Argentina, Belgium, Canada, Columbia, Denmark, Mexico, Nigeria, Norway, Poland, South Africa, Switzerland. I thank their Majesties the King and Queen of Sweden for giving me permission to study the menus for 11th December despite their private nature, and for allowing me to publish them.

My thanks to First Marshal of the Court Johan Fischerström and Surveyor of the Court Ingela Lilliehöök at the Royal Court for all the help I received concerning the missing menus for December 11 after 1950, for all the answers to my other questions and for their kind interest. I also thank Monica Friberg. who worked at the Royal Palace Household, for her great kindness in helping me to borrow photographs and obtain information, and for her interest in my project. My thanks to Court Photographer Alexis Daflos for his help with the Court menus and for his interest.

I give my thanks to Jan Brunius and Lars Wickström at the Palace Archives for their kind attention and helpfulness and for the great interest they showed in my ideas for this book. What would the December 11 menus for the years 1904-1950 have looked like without your help?

I thank Ulla Hager for her help in translating the French menus, mainly for December 11, although we do not know each other and have not even met. Your help has saved me a great deal of time. My thanks to Carl-Henrik Wendt for arranging the contact.

Many thanks to my fellow doctoral students at the Department of Economic History, Therese Nordlund, Sven Hellroth and Ursula Hård, for their moral and mental support in our daily toil, and to Therese in particular for all her proof-reading and for her penetrating comments on my rough drafts. I thank my colleague Jonathan Metzger for having the kindness to read and comment on the section that deals with Jewish food in Chapter 3, and I also thank Per Simonsson for all his technical computer help.

I am grateful to Björn Sandahl, historian and fellow doctoral student, for always believing in my idea (unlike some others) that it is possible to write about gastronomy from a historical perspective, and for having supervised me during my undergraduate studies at the university. Without your support neither my essays nor this book would have been written (not by me, at any rate).

My thanks are due to Tove Kjellander at the university library for her never-failing interest in my project and for our very rewarding email correspondence. I hope it will continue for many years to come.

I am most grateful to Torbjörn Boström, the photographer who worked with me in connection with photographing the Nobel menus for the years 1920 and 1945. I hope we shall continue to collaborate in the future. I also thank Gunnar Eriksson, head chef for the Golden Hall, Blue Hall and Stockhholm City Hall kitchens, for his interest and great help while photographing the above-mentioned menus.

My thanks to Holger Edström, who showed great generosity and kindness in allowing me to use the photos he took in 1991.

I thank the Sandviken School of Catering for two wonderful high-school years in the early 1980s. What creativity!

Many warm thanks to Trygve Carlsson, chief editor at Carlssons publishing house, for all his trusting confidence and enjoyable discussions and for believing in me and my ideas about this book.

I am grateful to Maria Asp at the Royal Academy of Sciences for her great help in locating for me the unique menu for 1907.

My thanks also to Anders Bárány at the Nobel Museum for the interest he showed in my work and for reading and commenting on all of it. That was worth its weight in gold.

Last, but definitely not least, I thank the Nobel Foundation and Lars-Göran Andersson for their enormous support and their great interest in my project. Believe me, that was worth a very great deal!

In my last conversation with Wilhelm, he expressed the hope and wish that I would manage to collect all the material for a book and that it would be an enthralling book. I leave it to you, my reader, to decide how enthralling it is.

Happy reading!

Ulrika Söderlind
Spånga, Sweden
February 2005

The Nobel Banquets

A Century of Culinary History
(1901–2001)

Chapter 1

The Nobel Prizes: Background

Alfred Nobel's funeral and his last will and testament

The funeral of Alfred Nobel was worthy of a prince. Thousands of people had gathered outside Stockholm Cathedral an hour before the actual funeral service was to begin. Nobel's relatives and friends gathered in the nave in a seemingly unceasing stream of guests. The flower arrangements were magnificent; from the outdoor entrance up to the altar there was an avenue of several lines of flowers. The nave pillars were covered with great, high foliage plants whose crowns reached right up to the ceiling. The nave itself was turned into a forest of palm trees. Two garlands of foliage formed a cross above the catafalque and a white dove hung from the vault, as if hovering above the coffin. The coffin was of light, polished wood and at one end there was a silver plaque inscribed quite simply: *Nobel, Alfred Bernhard, Born 21 October 1833, Died 10 December 1896.* The coffin itself seemed to float freely above the ocean of flowers below, but it was not covered with wreaths. Instead, the ocean of flowers consisted of wreaths from almost all corners of the world. At 3 o'clock, all the guests had arrived and the service began, with an introduction on the organ followed by a male choir. The Reverend Kjellman-Göransson then went over to the catafalque and performed the funeral rites. After the ceremony was over, all the wreaths were gathered together and Alfred Nobel's coffin was lifted up without any flowers and carried, to the accompaniment of a funeral march, to the awaiting hearse. The wreaths were carried on open carriages. At about half past four, the procession began. Stockholm had seldom seen such a cortège. Before the hearse rode torch bearers and in all the carriages in the cortège, which seemed never to come to an end, the coach lamps were burning. The people of Stockholm had turned out in force to see the cortège wind its way to the crematorium, forming two lines along the streets all the way up to Norrtull (the North Tollgate). A large number of torch bearers met the coffin outside the crematorium, where the lid was removed and an inner, white coffin was lifted out and carried in to be cremated. Alfred had expressly written in his will that his remains should be cremated, but this was not to take place until "after my death the arteries have been cut open and when this has been done and sure signs of death have been certified by competent doctors, my remains shall be cremated in a crematorium." Alfred Nobel's ashes were then buried at Norra begravningsplatsen (the North Cemetery) in Stockholm.

On December 14 in the same year Alfred Nobel died, his last will and testament was read. It was considered to be "most remarkable". Apart from a number of bequests to individuals such as relatives, close friends and servants, the will contained the following:

> "The whole of my remaining realisable estate shall be disposed of as follows: the capital, invested in safe securities by my executors, shall constitute a fund, the interest from which shall be distributed annually as prizes to those who during the preceding year have done most good for mankind. The said interest shall be divided into five equal parts, which shall be apportioned as follows: one part to the person who has made the most important discovery or invention in the field of physics; one part to the person who has made the most important discovery or improvement in the field of chemistry; one part to the person who has made the most important discovery in the field of physiology or medicine; one part to the person who in literature has produced the most excellent work of an idealistic nature: and one part to the person who has done most or best for fraternity between nations, for the abolition or reduction of standing armies and for holding and promoting peace conferences. The prizes for physics and chemistry shall be awarded by the Royal Swedish Academy of Sciences; that for physiological or medical work by the Karolinska Institute in Stockholm; that for literature by the Swedish Academy in Stockholm; and that for the champions of peace by a committee of five persons elected by the Norwegian Storting. It is

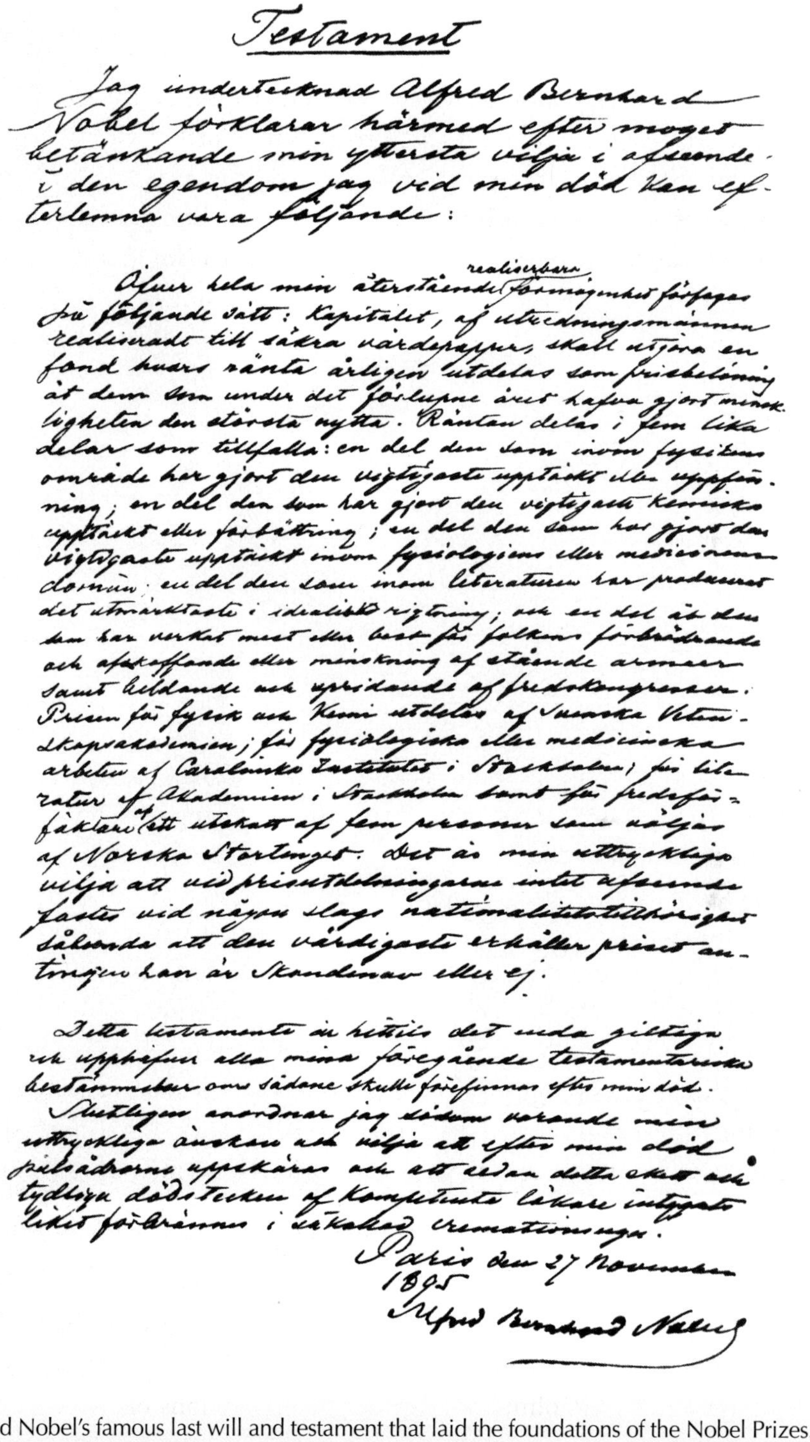

Testament

Jag undertecknad Alfred Bernhard Nobel förklarar härmed efter moget betänkande min yttersta vilja i afseende å den egendom jag vid min död kan efterlemna vara följande:

Öfver hela min återstående realiserbara förmögenhet förfogas på följande sätt: Kapitalet, af utredningsmännen realiseradt till säkra värdepapper, skall utgöra en fond hvars ränta årligen utdelas som prisbelöning åt dem som under det förlupne året hafva gjort menskligheten den största nytta. Räntan delas i fem lika delar som tillfalla: en del den som inom fysikens område har gjort den vigtigaste upptäckt eller uppfinning; en del den som har gjort den vigtigaste kemiska upptäckt eller förbättring; en del den som har gjort den vigtigaste upptäckt inom fysiologiens eller medicinens domän; en del den som inom literaturen har producerat det utmärktaste i idealisk rigtning; och en del åt den som har verkat mest eller best för folkens förbrödrande och afskaffande eller minskning af stående armeer samt bildande och spridande af fredskongresser. Prisen för fysik och kemi utdelas af Svenska Vetenskapsakademien; för fysiologiska eller medicinska arbeten af Carolinska Institutet i Stockholm; för literatur af Akademien i Stockholm samt för fredsförfäktare af ett utskott af fem personer som väljas af Norska Stortinget. Det är min uttryckliga vilja att vid prisutdelningarne intet afseende fästes vid någon slags nationalitetstillhörighet sålunda att den värdigaste erhåller priset antingen han är Skandinav eller ej.

Detta testamente är hittills det enda giltiga och upphäfver alla mina föregående testamentariska bestämmelser om sådane skulle förefinnas efter min död.

Slutligen anordnar jag såsom varande min uttryckliga önskan och vilja att efter min död pulsådrorna uppskäras och att sedan detta skett och tydliga dödstecken af kompetenta läkare intygats liket förbrännes i så kalladt crematorium.

Paris den 27 November 1895

Alfred Bernhard Nobel

Alfred Nobel's famous last will and testament that laid the foundations of the Nobel Prizes.
© The Nobel Foundation

Alfred Nobel as a young man © The Nobel Foundation

> my express wish that when awarding the prizes, no consideration shall be given to the nationality of the candidates but that the most worthy person shall receive the prize, whether he be a Scandinavian or not."

When the lawyer Mr Bo Löfgren (only 31 years old), who had been present when the will was read, came home to his wife, he told her: "Today I have seen something quite remarkable, something that will be of great significance."

The will attracted attention, not only in Sweden but also in other parts of the world. In London it created a sensation, and the Daily News considered that Alfred Nobel should be seen as a promoter of peace since he so strongly opposed war although as an inventor he had constructed something that had caused most devastation in the world so far.

It was not possible to estimate the value of Alfred Nobel's estate when the will was read, but it would not be wrong to guess that his assets amounted to between 30 and 35 million kronor at his death. If the rate of interest at that time was reckoned to be not more than 3 per cent, the amount of money that could be apportioned to the above-mentioned five subjects was about 900,000 kronor — in other words, a great fortune every year. People wrote about Alfred Nobel's will and it was considered "a gift to mankind". This gift to mankind was first awarded in 1901 on the day of Nobel's death, after much legal quibbling. The ceremony in Sweden was held in the Main Hall of the Academy of Music in Stockholm at 7 p.m. The winners of the Peace Prize were announced in Kristiania, Norway, by Stortinget, and at the same time telegrams were sent to the Norwegian king and the prize winners. None of them were present at the ceremony. The prizes were sent immediately to the prize winners and the medals and diplomas later. The Norwegian ceremony was very simple, taking only about 10 minutes.

Chapter 2

Gastronomic Man

Eating and drinking are often described as activities that people practise when they need to get the life-supporting supply of nourishment that is necessary for survival. Eating and drinking are basic, right at the bottom of Maslow's hierarchy of needs. According to this model, people must first satisfy these needs before devoting energy to satisfying other needs that are higher up in the hierarchy. But are food and drink really as simple as that? In the following, I shall analyse the factors that I consider to be very important and decisive for what, when and how people eat and drink what they do.

Let us begin by looking at the meaning of words that are connected with food and drink. The word diet, which is absolutely central in this chapter, means a combination of the different kinds of foods that are consumed daily to meet a person's need for nourishment. The form of the diet can vary but should be sufficient for nutritional needs, provided the various foods are combined in a suitable way. The word diet comes from the Greek word *diaita*, meaning way of life or lifestyle. Today it has come to mean a special combination of foodstuffs, consisting usually of protein, carbohydrates, fat, vitamins, minerals, water and fibre, or roughage. Every diet should be suited to a person's age, sex and physical activity. For various reasons, some people need larger quantities of some of these components than others, or should avoid some of them altogether. Our diet consists of various kinds of foodstuffs, products that are used to make our food and drink. Generally speaking, food is divided into two main groups: vegetables, which come from the plant kingdom, and meat, which comes from the animal kingdom. Some of these foodstuffs can be eaten without cooking, such as fruit and certain vegetables, while others need to be prepared by, for example, being baked, roasted, grilled or boiled. Now that I have come to the preparation of food, it is time to talk about the art of cooking which, in Sweden at least, has come to mean the technique and art of preparing food. In turn, the word food means what has been prepared for eating by human beings. A term we often come across in connection with food is the word gastronomy. This word comes from the Greek *gastronomia.* meaning knowledge of the stomach. It came into English via the French gastronomie, which was first used in France in the early 19th century. In modern usage, this word has come to mean the science of high-class cooking. Now that we have completed the list of central concepts and terms, it is time to consider the factors that steer our choice of food and drink.

Needs

Human beings are by nature physical beings and their bodies are dependent on nutrition or food of a certain composition. The supply has to be provided fairly regularly because their ability to store food is limited. Their working capabilities depend on both physique and the supply of food.

Edibility

The term eating habits holds the answer to the question why people eat what they do. Habits in themselves mean that certain food is so self-evident that people eat it continually, while rejecting that which they find foreign, or eating it with caution. Individual eating habits do not have much to do with the individual's personal taste; they are shared with those of the members of the family and the people living in the neighbourhood. The repetitive element in human eating habits is not restricted to a person's private life; it belongs to the cultural heritage of mankind and is remarkably stable. Studies of eating habits carried out in the US among Scandinavian immigrants show that they are faithful to their Scandinavian Christmas fare even in the third and fourth generation, long after they have stopped speaking their native language.

An important factor in the edibility of food is the technology of all the different ways of getting food: hunting, fishing, cattle farming, agriculture, industry and so on. If this broad term is applied to a more local level in the community, it becomes more limited, comprising food preparation techniques such as preserving, butter-making, cheese-making, baking, brewing, distilling and so on. These techniques are very ancient as well as being essential for life since many of nature's resources will only keep for a limited time in their natural state.

Ideology is another factor affecting the edibility of foodstuffs for human beings. An ideology in itself is governed by various factors like religion, medical concepts or moral concepts. In religion, it is the rites that are the visible or practical aspect of the ideology that govern to a great extent what is considered edible or not; such rites often originate in religious writings. An interest in eating habits based on medical aspects existed even in antiquity. A similar interest based on moral aspects is found in the temperance movement, which classifies wine and spirits as poisons, and not as a source of pleasure.

Availability

The availability of raw materials and foodstuffs also plays an important role when people choose what food to eat. In turn, availability is dependent on factors like nearness, economics and regulations. The geographical location of different foodstuffs is important for what ends up on our plates. Foodstuffs that have been transported over long distances are naturally more expensive than local products, which is where the economic factor comes in. Our choice of foodstuffs and raw materials is connected with our own financial state. Finally, regulations also affect the availability of food; the food rationing system introduced during the world wars exemplifies this factor.

Senses

Food appeals to at least four of our five senses. The first thing people eat with is their eyes, which is why food should look good and appetising.

The smell of food can arouse feelings of desire when you are hungry and feelings of displeasure when you are full up or if the food smells nasty.

Beverages like wine can also be enjoyed via their bouquet at the beginning of a meal.

Our sense of feeling tells us whether food is hot, cold or tepid. When we chew food, we find out whether it is hard, soft, tough or tender. All these variants play a part in the right context.

Our sense of taste tells us about the inner qualities of the food: whether it is sweet, salt, bitter or sour, mild or strong. All these variants are also relevant, but only in a certain order. Salt food must come first and sweet food last.

Gastronomic reflections

Anthelme Brillat-Savarin published at the end of his life *Physiologie du goût* (The Physiology of Taste). Brillat-Savarin's book is not a pure cookery book; in it, he discusses and philosophises on culinary art and its meaning. He tells the reader that his book has two purposes. The first is to discuss the basic theories of gastronomy and the second to defend gourmands from the reputation they have for being greedy omnivores without any finesse at all. This book is considered important in culinary literature because it was written by a man who did not see himself as a self-appointed food expert, yet it is written stylishly and lucidly, revealing a genuine interest in food. According to Brillat-Savarin, gastronomy is an organised science that is related to other sciences such as:

— Natural science, because of its classification of nutritious substances.
— Physics, because of its investigation of composition and properties.
— Chemistry, because of the analyses and solutions it provides.
— Culinary science, because it deals with the art of preparing food and making it enjoyable to the taste.
— Commerce, because of its discovery of methods for purchasing the materials used as cheaply as possible and disposing of what is for sale as advantageously as possible.
— Economics, because of the sources of income it provides for the state and the ways of exchanging goods that it opens up for people.

Over and above all that, gastronomy dominates people's lives from the cradle to the grave, as well as affecting all the social classes. The material objects of gastronomy are everything that can be eaten. The first aim of gastronomy is to keep the individual on his or her feet.

Gastronomy as a science

Physiological taste research

Perceptions of taste are aroused only when special sensory organs in the tongue, palate and throat are stimulated. Because of their appearance under the microscope, these organs are called taste buds. They consist of large, oblong cells that have protoplasmic shoots directed towards the narrow, outer taste pores. These taste cells and their fine shoots have a relatively short life — only a couple of weeks. They are replaced continuously as undifferentiated epithelial cells surrounding the taste buds develop. When these simple cells touch the specific taste nerve fibres, they are transformed into taste cells, with their microvilli and special response patterns to various kinds of stimulation. If the taste nerves are cut off, all the taste buds in the nerves' area of distribution on the tongue disappear and return only when the taste nerve fibres on the tongue have regrown.

As for the signals in the taste nerves, the taste nerve fibres run from the front three-quarters of the tongue in the nerve's large sensory nerve. However, these fibres differ from other nerve fibres in the thick nerve trunk, forming a thin nerve which, after a short distance, runs through its own bone canal in the middle ear and passes the anvil, after which it joins the facial motor nerve and passes through that nerve's bone canal to the brainstem. Communication between the taste buds and the brain takes place via nerve fibres that send a very simple code to the brain — it has only one signal and it reaches the brain without any changes or interference. There are, however, very great individual variations in taste; that is, the way taste is perceived differs from one person to another.

The fact that eating is not merely something physiological but also to a high degree a social act makes food and drink an intermediary of social relations. One and the same dish will not taste differently just because it has been prepared differently but because its special taste is also determined by its social context, that is, who prepared the food, who we eat the food with and the situation in which we eat the meal.

Recent research has shown that, in addition to the four basic tastes, salt, sour, sweet and bitter, humans can identify a fifth taste. The Japanese have known about this fifth basic taste for a long time and call it "unami", meaning delicious, piquant. This fifth taste has to do with the taste of amino acids, for example monosodium glutamate (MSG). MSG has been used to reinforce tastes in Asia for centuries and is a normal additive in soups, for example. Scientists have now succeeded in finding receptors on the tongue that react to amino acids, that is, unami, and that this basic taste plays a large part in the way people appreciate wine.

The physiology of thirst

When you have a deficit of water in your body, you feel thirsty, which creates a need to drink. If this is not heeded in time, it can lead to disaster. You can go without food for weeks, but without liquids you die after only a few days. Thirst is included among the so-called general feelings. The desire to drink is triggered by a limited area in the middle brain, where there are special nerves that are sensitive to dehydration. When the water content in the blood falls, these nerves send a signal to the corpus callosum, which results in a conscious need to drink; the corpus callosum plays a vital part here, since it has been shown that animals without a cerebrum do not drink spontaneously. Only when the body's need of liquids is satisfied does thirst disappear; it is also possible to quench burning thirst very temporarily by drinking a liquid, but if the need of liquids in the body is not satisfied, you soon feel thirsty again. This means that thirst is inhibited in two principally different ways. The act of drinking itself triggers nerve impulses in various sensory nerves in the oral cavity, throat and stomach. These impulses inhibit the feeling of thirst temporarily long before the liquid has had time to be absorbed and affect the blood's osmotic pressure. Not until enough liquid has been absorbed for the blood's water content to return to normal does the feeling of thirst normally disappear.

Sociological gastronomy

Food culture

It is a physiological and biological fact that all living creatures have to eat and drink to survive.

An early photograph in the history of the Nobel Banquets. It was taken in 1925 and shows the Hall of Mirrors at the Grand Hotel, Stockholm. The tables are arranged in a horseshoe shape, with tables in the centre to fill the hall. © Pressens Bild

Another physiological fact is that a creature cannot eat or drink just anything. Certain organisms or species can assimilate a broad spectrum of foodstuffs while others can only live on one type of food. Human beings can, in a physiological sense, assimilate a large number of foodstuffs, which makes us omnivorous. Nevertheless, there is a great deal that the human body does not tolerate; for example, no one can drink a litre of 96 percent alcohol and survive. Thus there are physiological limitations to what can be used as human food and what cannot be used. These limits are set by nature.

Such limits are broader than those set by human culture. Every society selects a part of what, according to nature, can be considered human food. This cultural definition of what can or cannot be considered food (that is, food and drink) is one of society's fundamental norms. These norms or imprinting are something that every person is exposed to from a very early age. The culturally defined foodstuffs vary from one society to another, so food is part of a country's cultural character. The food culture of a country or a people is not uniform but has regional variations. These variations can be based on geography and/or economics. Societies have for a long time had two parallel cuisines. Usually one of them belongs to the broad majority of the population. From a historical point of view, this cuisine only changed marginally over the years until the arrival of refrigerating and freezing technology. The other cuisine used to belong to the elite of society — the upper classes living both in towns and in the country. This cuisine has been more receptive to variations in what is eaten and the influences have often come from abroad. The eating habits of the upper classes gradually seeped down to the cuisine of the broad majority, which in turn meant that the upper classes adopted new food habits since the previous ones were no longer reserved for them. As if it was not enough for there to be parallel cuisines for different social classes, people have also divided up the dishes that can be prepared in their kitchens into everyday food and festive food. Festivals can be of a sad kind, such as a funeral, or of a happy kind, such as a birthday. The different ways in which meals are prepared can be traced far back in time; archaeological excavations in Sweden have traced them as far back as to the early Middle Ages. However, this is presumably not unique for Sweden but rather a global phenomenon. It will be interesting to see if future investigations can prove the existence of this phenomenon in even earlier times.

Food cultures are created not only by geographical conditions but also by consumption, or perhaps the other way round, that everything people eat and drink appears as consumption. Thoughts about consumption go far back in history and have changed with time, but there are certain concepts that have always been there. One of them is that people's needs expand and our wish to meet this expanding need drives economic development. This means that consumption in itself is a kind of arena for social competition and identity formation. In addition, the coveted value of more exclusive consumption runs the continual risk of being weakened by people's search for distinguishing markers.

Food culture may be poor or rich, and consist of a very few dishes or an abundance of them. Regardless of poverty or abundance, there are very strict rules/norms in all cultures about what can be considered a dish or a food element. This division into what can be eaten and what cannot be eaten is both fundamental and extremely emotionally charged. An example is that during famine years in Sweden (historically speaking) people did not eat edible fungi because people in the society of that time did not consider that fungi had any connection at all with food. People starved rather than eat fungi. However, food culture does not consist merely of a number of food elements; it not only has limited boundaries but also has strict rules within the culture itself as to what food elements can be combined only in a certain way to be counted as a dish. For example, in Sweden, meatballs, lingonberry jam, sauce and potatoes are an accepted combination. Recipes are the grammar of food. The fact that food forms part of culture and is not just part of nature is a specific feature of the human race. Mankind's ability to move from being a nature child to being civilised is expressed in its ability to modify the world according to its own intentions. Regarding food, the move from nature to culture takes place when people prepare their food, cultivating and processing it. Food acquires a cultural hallmark with the aid of various utensils like pots and pans and the like. Making food is a cultural act in which a connection occurs between the food and the person preparing it. In turn, a prepared meal is a dialogue between that person and the person who eats it. Meals/dishes are thus a form of communication in which the eater eats more than what is on the plate.

Because our relationship to food is formed early in life (when we are still infants), tastes are people's most conservative preferences. Eating habits are the last ones we change. This becomes very evident when we move from one country to another. We learn the new country's language and customs but we prefer to eat the food we grew up with. Eating habits are something that develops in a socioeconomic and cultural environment and are what it is easiest to take from our own culture when we move to a new country. In the big cities of southern countries, the meal schedule in the summer often consists of an almost nonexistent breakfast, a small lunch, and a largish snack of biscuits, sandwiches and the like eaten around 6–7 p.m. The big dinner is eaten after 9 p.m. In the countryside in these countries, the day begins very early with a big breakfast, and during the day there are a large number of snacks before the evening, when the main meal of the day is eaten. When an immigrant meets the new country's customs and eating habits, his or her eating habits change on two levels. The first level involves changing the traditional combination of foodstuffs; in other words, the gastronomy itself changes. Because new foodstuffs are incorporated into the diet, the meals are given a different nutritional content than before. Level two is connected with the mealtimes. With some delay, the immigrant starts to choose new foodstuffs and the mealtimes gradually begin to agree with the times at which the host country eats.

Food is classified as staple food and supplementary food. Staple food is, for example, bread, potatoes, rice, pasta and the like, while supplementary food is, for example, various kinds of meat, dairy products and vegetables. The gastronomic tradition of every culture comprises and is defined by its foodstuffs. These are seen to be the "real" foodstuffs that the people have a strong emotional bond with, since they are the ones they know best. Since there is this strong bond between the people and their foodstuffs, they are only to a very limited extent affected by migration. People's bonds to supplementary food are not equally strong as they are not so necessary for the survival and continuance of the traditional eating habits. Supplementary food is usually used to reinforce the taste of the staple food. This group of foodstuffs is influenced almost immediately after migration as people do not have such a strong bond with them. Studies have shown that the consumption of sweet products, new sources of fat, snacks, drinks, ice-cream and so on increases among immigrants, who otherwise continue to choose the same staple foods and the same kinds of meat, dairy products and vegetables as they used in their home countries.

This means that the eating habits of immigrants change in a fixed order, within which cultural/psychological bonding and reinforced taste form two extreme poles along a continuum. When a traditional meal is to be made in the new country, the choice of foodstuffs swings from the cultural/psychological pole to the pole that concerns taste. Staple foods that are known are retained as they provide a sense of security in a foreign environment. When new foodstuffs are added to the diet, the selection mechanism moves in the other direction; the aspect of taste is given priority. Immigrants introduce the host country's most tasty foodstuffs first, like sweets, nuts and new sources of fat. After sweets and fats have been incorporated, foodstuffs that have a stronger cultural bond are selected, like meat and dairy products. Staple foods are those that are changed last, if ever. Immigrants' adaptation of their diet to the new culture is total and complete only when the immigrant population completely accepts the new country's staple foods. However, immigrant groups retain their staple foods for several generations, so total adaptation to a new diet takes a very long time, even up to 100 years.

However, it is not just eating habits that change after migration; so do meal times. This is because, when the pattern of work and leisure time changes, meal times also change for immigrants so that they fit in with the pattern that exists in the new host country. Here, too, a feeling of security and identification with their culture plays an important and strong role in immigrant groups. Thus the meal that is most strongly charged with emotion is changed last, while the meals that are more neutral or connected with changes in taste are changed first. The first change is that the eating of snacks between meals and the consumption of soft drinks increases rapidly among newly arrived immigrants (with negative consequences for their teeth). The second change occurs in the meal that is considered least culturally significant, namely breakfast. Immigrants come into contact with the host country's eating traditions through this meal. The evening meal or the occasion when the family get together is reserved in order to strengthen the feelings of family and security in each of the family members. In other words, immigrant families prefer their home country's food for dinner, and this meal changes last. Looking at the way the diet changes during the week, it is the working days that are affected first, while Sundays are reserved for the eating habits of the home country. The food eaten at festivals is that which changes absolutely last.

Religious beliefs and meals

Festive meals in a country are often governed by ancient religious beliefs and traditions. In this respect, religion stands for an ideology that controls the choice of food at these festivals. In Sweden, this is exemplified by the eating of "semlor" (Lenten cream buns) before Easter. This tradition dates back to the Catholic celebration of Easter, beginning with 40 days of fasting during Lent, preceded by Shrove Tuesday, which is the last day before Lent. This was the day of feasting when people gorged themselves with sweetmeats of various kinds before the fasting started. Fasting before festivals takes place/has taken place in many religions. Religious fasting seldom involves total abstinence from food. What usually happens is that the practising individual eats fewer calories than the normal intake; in other words, people abstain from meat, fat and sweets. There is, however, a stricter form of religious fasting that practises abstention from all animal products, sometimes called ascetic fasting. Very often religion is more than just a faith in God for religious people; it permeates their whole existence so that they conduct their life according to the rules of the religion. Most religions have rules about how people should and must deal with their food. These rules regulate what can be eaten and what cannot be eaten, how animals should be treated when they are slaughtered and how foodstuffs should be handled and prepared.

Even those people who were not members of an established religion but believed in so-called folklore were governed by these eating rules. Belief in fate plays a central role in folklore, which means that the ideas that control eating can be seen as a way of appeasing or controlling fate. Some researchers claim that the custom of taking a gift with you when you are invited to a family is based on the idea that it was harmful to come empty-handed to a woman who was about to give birth. The food that was taken to women who were lying in may be seen as an offering to the powers of birth and fate. After a child had been christened, it had to be given bread so that bread would never be lacking in the child's future life. That is why bread was also placed in its swaddling clothes. If a child had been weaned, it was not allowed to breastfeed again since it would then run the serious risk of becoming a child that brought bad luck. In addition, it was considered that what a child ate could shape its character. If its mother ate and drank a lot while she was breastfeeding, the child would become a glutton and have a big appetite later in life. If it ate off a knife, it would learn to be a thief. If it ate pork from the pig's rump, it would stop growing, and if it ate hare's heart, it would be timid. Children should eat up everything on their plate; the last bits were called the fun bits or the power bits. It was also believed that if these fun or power bits fell into the wrong hands, the child could come to harm, because it was thought that the remains on the plate were a part of the child itself. If the whole plate was eaten up, the last cake on the cake dish could not be eaten because then the family would end up in poverty.

Table manners, seating lists and tableware

In the late 19th century and at the turn of the century, table manners were the very symbol of the cultivated person, in which the use of cutlery played a very central role. You were not allowed to eat any dish with the hands only, not even a sandwich. However, the use of cutlery is not the only sign of cultivation at the dining table. Table manners as such form a complete institution that trains the whole body and mealtimes as inter-personal operations. You must not sit in any old way on your chair but at the right angle and facing straight forward; you must not lean on the table with your arms or elbows; you must not chew with your mouth open; and you must not slurp or smack your lips.

Table seating in most cultures has usually shown the guests' relative rank. During the Renaissance and the Baroque Period, for example, royalty sat at their own table. When King Gustav III of Sweden travelled to Italy, he was accompanied by the not yet ennobled Johan Tobias Sergel, who had to eat with the servants, because only the nobility were allowed to sit at the king's table. In peasant homes, men and women sat on opposite sides of the table; in fact, it was often only the men who sat down to table to eat, while the women stood and ate. The master's self-evident place was at the head of the table. The seating of the guests of honour at large dinners has changed over the years; for many years there was an upper and finer end and a lower end for the less important guests. This system changed in the 19th century, however, when the host and hostess sat opposite each other at the middle of the table with the guests of honour around them.

Tablecloths were introduced in the Middle Ages at the royal court and in the homes of aristocrats, clergymen and rich merchants. Laying the table was such an important part of a feast that special table-layers were needed. In the 15th century, the table was covered with several patterned cloths placed on top of each other. The bottom one was a thick cloth, a piece of tapestry or gold brocade that reached right down to the floor on all sides. On top of this was a white cloth that had been mangled so that a square pattern was created. On top of this cloth yet another white cloth was laid that was shorter than the other one and mangled with another pattern. The top cloth was changed between each course and was also used as a napkin for the guests to wipe their hands and mouth on. At the court of Erik XIV in 1562, long napkins were introduced, which were later replaced by individual napkins. The members of the Swedish royal family still have an extra small cloth placed under their covers, a reminder that those who sat at the same table could underline the difference in rank by not eating on the same cloth. When peasants were visited by the priest, a special priest's cloth was placed on the table under his cover. At a peasant family's table, the difference in rank between the guests could be emphasised by the host and the guests eating at a table with a cloth on it while the end of the table where the servants sat had no cloth. Until the 1930s, white damask was in principle the most usual type of tablecloth in most social classes, but then coloured linen cloth with simply folded napkins became fashionable. Since the Second World War, more brightly coloured tablecloths have become the rule.

A long napkin was a cloth that usually ran the whole length of the table. It was placed so that the plates lay on one half while the other half was folded over them, so that they were not be dirtied by things falling down from the ceiling — which often had holes in it — even though the table was laid several days before the dinner. When the guests came to the table, the upper half of the cloth was folded down over their knees and used as a napkin. The hand towel that was used during the Middle Ages for drying your hands after washing them before the meal can be seen as a predecessor to the individual napkin. In the mid-16th century, individual napkins became prevalent on the continent, and in Sweden they were first

Crown Princess Victoria's cover at the Nobel Banquet at the Royal Palace on December 11, 1995. Note the eggcup in the top left-hand corner of the cover. This eggcup belonged to King Karl XIV Johan of Sweden. In his opinion, there should always be an eggcup and salt on the table so that guests did not need to leave the table hungry if they did not like the food that was served. Moreover, it was very difficult to put poison in a boiled, unopened egg. © Kungl. Husgerådskammaren. Photographer: Alexis Daflos used at the court of Erik XIV; in 1562 he had 150 such napkins. To begin with, the napkin was laid across the left arm or shoulder, but when clothes with large ruffles came into fashion, it was tied round the neck. In the 18th century, the gentlemen used to tuck it into their waistcoat while the ladies used a pin or brooch to fasten it to one shoulder or in their décolletage. This custom continued into the 19th century, but in the 20th century napkins were placed over the knees. In the 19th century, a napkin was about 80 cm square but in the 20th century it decreased to about 50 cm. The way napkins were folded has always been an important part of their use. This custom was introduced in the 16th century when forks gave napkins a new role/function as a status symbol instead of just being a necessary wiping cloth. Napkin rings and napkin holders started to appear around the 1840s. These were used mainly in middle-class homes in towns, while the upper classes used their napkins only once before they were washed, and labourers and peasants did not use them at all. There are napkins preserved in Sweden that were used in peasant homes, but they were used as shawls or for holding the prayer book when the peasants went to church. Rich farmers began to use napkins in the 1870s in Sweden, but these were mostly used for decoration under the bridal couple's or the priest's plates at weddings. When the guests in these circles started to be provided with napkins, they often put them aside so as not to trouble the hostess with the need to wash them.

In the Middle Ages, it was customary to use a slice of bread as a plate and the meal was rounded off by eating it. There were also rectangular wooden dishes that two or three people could put on their knees and eat off together. It was from these wooden dishes that both individual wooden plates (first square and later round) and table tops with a frame developed. The development from bread plates to metal or wooden dishes took place in the 15th century among the upper classes. During the reign of Erik XIV, wooden dishes disappeared from the Swedish court and were replaced by pewter or silver platters. In the 17th century, the new aristocracy could afford silver plates. Pewter continued to be the material used by the upper- middle class well into the 18th century and as late as the 17th century the lesser nobility were still using wooden plates for everyday use. During the 17th century, Sweden started to import Dutch plates made of glazed earthenware, and these became common among the upper classes during the 18th century. The difference between deep plates and flat plates emerged during the 17th century. In 1726, Sweden started its own ceramic industry when Rörstrand opened its glazed earthenware factory. In the 18th century, porcelain plates also started to be imported into Sweden thanks to the Swedish East India Company. Not long after, Sweden began its own production of porcelain.

The spoon is the oldest eating utensil we know. A find of pottery spoons was made in Sweden that have been dated to the Stone Age. Silver spoons were first made in Sweden in the Middle Ages. The farmers and merchants who bought these spoons did so not to have them for everyday use but to use them as schnapps spoons at big dinners; or perhaps they served as an investment and status symbol. Those who could not afford silver spoons made do with similarly shaped spoons made of brass. In the 16th century, you often took your own spoon with you in a small case or in your pocket when you went to a party. Silver spoons have changed in shape over the course of time; the earliest ones had a round bowl and a short, often cylindrical handle, topped with a large knob or, in the Middle Ages, the figure of a saint. Towards the end of the 17th century, a new type was introduced, the rat-tail spoon. This had a longer and flatter handle which was joined to an egg-shaped bowl in a characteristic way: The handle was pulled to a point which was soldered to the underside of the bowl. To begin with, this joint continued a long way under the bowl, looking like a rat's tail, but during the 18th century it shrank and disappeared. When, during the 18th century, spoons and other eating utensils started to be made by pressing metal, a rich variety of models developed.

The knife was introduced as a part of table cutlery towards the end of the Middle Ages, but it was not common until the 17th century, and to start with only at court and among the aristocracy, when it began to be included in the case that guests took to meals for a long time, table knives had a blade of iron while the handle was of silver or bone of some sort, often with enamel inlays. Among the upper classes, mass-produced table knives became common in the 18th century. The oldest type of table knife, like other knives, had a point so that food could be speared, but this became unnecessary when the fork came into use.

Table forks gradually started to come into use at the courts of the northern Italian principalities and republics in the 14th and 15th centuries, but they were not a natural part of tableware until the 16th century, when they reached Sweden at the court of Erik XIV. Forks were usually placed alongside a table knife in the elegant cutlery cases that upper-class people took to parties. The lower classes were still without forks. In Germany, for example, there was some opposition to this new luxury item in the 16th century because superstitious people connected them with bad luck and black magic, probably on account of their pointed prongs (everything sharp was dangerous in folklore). It was not only impolite to point at someone with your fork; you might also in some magical way harm the person you pointed at, and the clink of a fork could summon up the devil. In the aristocracy and the upper-middle class, forks spread as eating utensils during the 17th century, but it was not until the 19th century that they became common among peasants and labourers. The number of prongs on a fork has changed over the years. During the Renaissance and the Baroque Period, they had two steel prongs and a handle of bone that was often inlaid with or made of silver, often gilded. In the 18th century, the forks made of pressed silver had three prongs and in the 19th century came the fourth prong. When, during the 19th century, the lower classes started to use forks for eating, they were often made of steel and had a wooden handle. Marine-archaeological finds from the warship Kronan, which sank off Öland in 1676, show that there were four-pronged forks in use on Swedish ships as early as that.

Drinking vessels, like spoons, are very old and a self-evident part of every cultural inventory. Drinking vessels were first made as status symbols and for ceremonial use; in everyday life, people made do for a long time with what nature could offer — mussel shells and gourds, for example. During the Renaissance, drinking horns were inlaid with precious metals. Wood, metal, glass and ceramics were the materials used for a large number of basic designs. The wide, open, wooden drinking bowl appeared in Scandinavia during the Viking era, and this type survived for many years among Swedish peasants; you took up your drink from the bowl with a cup or a ladle. The cylindrical or slightly conical goblet became common in Sweden at the end of the Middle Ages and was made of pewter, silver or even wood. Those made of wood were shaped on a lathe and were very common in the 16th century. Glass goblets were imported from the period of the Great Migration up to the 16th century. During the 16th century, these glass goblets became cheap mass goods; 35,000 of them were smashed at Kristian IV's coronation celebrations. Ceramic mugs or cups with a handle came much later. The earliest coffee services had no handle and became popular in the mid-18th century. Goblets with a high foot were typical of the Middle Ages and could also be made of wood. Goblets with a lid and handle became common during the Renaissance. Tankards with a handle and a lid were originally serving vessels but developed into individual drinking vessels that were very popular in the Baroque Period. These tankards were copied by the peasantry in wood. Ceramic tankards were first made in western Germany in the 16th century and were used typically in taverns until glass and porcelain had their breakthrough in the 18th century.

Schnapps cups were small drinking vessels made of silver, pewter or wood, hemispherical in shape but with a handle and a foot. They looked most like coffee cups with a handle. These cups are mentioned in written records in the late 17th century and were used for drinking aquavit. Marine-archaeological finds from the warship Vasa show that such schnapps cups were in use earlier than the late 17th century. The Vasa sank in 1628 and among the artefacts found on board were schnapps cups. A tumbler was also a small, hemispherical cup for aquavit, made of silver, pewter, brass or glass. It was constructed so that it returned to an upright position if it was tipped over. It often had a silver coin in the bottom and was popular in the 18th and 19th centuries. Yet another vessel that was made for drinking aquavit was the "öronskål", a little bowl with a handle. These were usually made of silver. A "snibbskål" was a round drinking vessel with four lips, that it, four corners just under the rim of the bowl. These bowls were often painted red and are known to have existed as early as the 16th century.

Meals as events

It is rather a complicated matter to speak about meals as events since different people interpret an event in different ways. Nevertheless, there are certain factors that those who arrange a meal utilise in order to create an experience from and round the meal. The factors they work with are colour, light, sound and design.

Colour

Colour is something that all people perceive differently because the eye sees light in different ways. When you see a colour, it is the result of different light wavelengths. The eye perceives these as different colours depending on the wavelength. Since all people perceive colours differently, a standard colour language has been developed. The system most commonly used today is the Swedish Natural Colour System (NCS). This uses six elementary colours: yellow, red, blue, green, white and black. The system describes a specific shade and the elementary colours are codes that function when it is necessary to denote an exact shade of colour. In chromatics, one speaks of hot and cold colours. Red and yellow are hot colours, blue and green are cold colours. There are objections, however, to calling red a hot colour as it is often experienced as cold, depending on the context. Chromatics says that the hot colours lie between the yellow and the red elementary colours in the colour circle and the cold colours between blue and green. Hot colours are experienced as energising and cold ones as soothing. When combining colours, it is important to be aware that colour setting can have unexpected consequences and that the colours take on a different shade than was intended. If this phenomenon is not taken into account, there may be unfortunate results, for example when putting together a menu for a restaurant. Anyone creating a mealtime experience ought to think in terms of colours so that it is an experience for all the senses. With the help of the right colours, it is possible to enhance the experience planned for the guests and to create an ambience around the meal itself.

Red is perceived both as a stimulating colour and as a sensuous colour but it can also be seen as a warning colour. Red in combination with yellow is thought to increase the guests' appetites, whereas red food alone is thought to give them more energy and greater vitality. Some people think that, to avoid giving the guests high blood pressure, increased muscular strength and strong reactions, meals should not be served in a room that is furnished totally in red.

Yellow is considered to be the colour of the sun, which is felt to be warm and hospitable. It is also thought to raise the level of alertness and to sharpen people's intelligence. In combination with black, yellow gives a warning.

Green is the colour of hope, of spring and of growth. It creates a feeling of calm and security. Green is felt to be pleasant.

Blue is the symbol of immortality. It is cool and creates calm as well as being emotionally stimulating. Blue can also give a feeling of coldness, making people think of shade. Thus a good composition of colours with well-balanced harmonies and colour chords for sight, hearing, taste and smell can enhance the guests' impressions, moods and experiences of the meal.

Lighting the candles on the tables before the centenary of the Nobel Banquets in 2001.
© The Nobel Foundation/Boo Jonsson

Light

Light is much more than just decoration; it affects the way a room is used and how an atmosphere is created when there is no daylight. Illumination affects the way we feel and is of great importance for the way we experience our surroundings as well as for the overall impression we get. Thus there are two different sources of light, artificial light and daylight. Choosing the right source of light for an environment is decisive for the success of the lighting. The choice of light also determines the character and quality of a room. There is a belief that the best illumination/source of light is that which cannot be directly observed; in other words, the source of light should not be noticed, it should only create a certain atmosphere in the room. People's attention is drawn to bright sources of light and these create a feeling of pleasure and have an energising effect; strong-coloured and moving illumination is felt to be unpleasant. When choosing the lighting for a room in which meals are to be taken, it is important to start with the room's function, that is, ceiling, walls, shapes and materials. Outdoor lighting is also important as it follows the guests all the way until they are seated at the table.

Sound

Since sound consists of mechanical vibrations in the air that the ear perceives, there are two sorts of sound that people hear: what is called music, that is, regular oscillations, and what is called noise, that is, irregular oscillations. A restaurant where music is played at a high tempo gets the staff to provide swifter service; this effect is part of the concept of fast-food chains. It is vital that the volume of the music does not drown the guests' conversation. The acoustics of a room are important since poor acoustics will disturb the guests and affect their overall enjoyment of the meal. The sound that reaches the human ear consists in part of words, but the rest is reflections from walls, floor, ceiling and furnishings. When these reflections do not reach the ear at the same time, acoustic problems occur.

Design

The guests' behaviour is influenced by the design of the room where the meal is eaten. Meals have become a social and cultural meeting place, which means that the meal and art go hand in hand. The combination of form, colour and lighting makes a whole that communicates feelings. Thus the choice of materials, furniture, porcelain, glass and cutlery is important as they are what govern how the guests perceive the room with their senses. Just as there are hot and cold colours, there are hot and cold materials; wood is felt to be warm while stainless steel is felt to be cold. How the porcelain, glass and cutlery used at a meal are designed plays a major part in the total meal experience. In other words, the cutlery has to function, both from an ergonomic view and as emotional expression; the right sort of glass has to be used for the right drink and so on.

Menus

The French word *menu* literally means "little table", but this "little table" has come to mean the bill of fare in a restaurant. The phrase *à la carte* on a menu literally means "according to the card" and describes the dishes that can be ordered. It is a list of all the dishes available with the price given for each one. A guest who chooses from the à la carte menu is free to compose his or her own meal by selecting dishes from this menu. An à la carte menu is often complementary to the day's menu, offering the guests greater variation, but the dishes are usually rather more expensive than those on the day's menu. An à la carte menu is normally composed and printed to last a long time, listing the starters, main dishes and desserts that the cuisine can offer. This type of menu can be very extensive at large restaurants, catering for almost every kind of taste.

The phrase *plat du jour* means today's special, often at a low price and served quickly. Having this kind of dish on the menu makes work in the kitchen more rational and effective and the chef can utilise what is on the market that particular day. *Prix fixe* on the menu means that there is a combination of dishes to be had at a fixed price. This type of menu is most common on Sundays and for dinner-dances. A menu with the word *arrangé* printed on it tells the guests that that the meal is one that has been ordered. Such specially arranged meals are ordered in advance from the head waiter. They can take various forms, but the simplest type consists of a starter and a main course, which can be added to with other dishes and drinks.

The arrangement of the dishes on the menu should be as simple as possible, clear, systematic and easy to read. The number of courses on a menu may vary, but regardless of the number, they should come in the following order: 1) cold starters, 2) hot starters, 3) soups, 4) egg dishes,

5) fish dishes, 6) meat dishes, 7) poultry dishes, 8) pastry dishes, 9) vegetable dishes, 10) desserts. If there are several fish, meat or poultry dishes on a menu, they should follow the golden rule: boiled before roast, whole before minced and light before dark.

The descriptions above show that eating and drinking is quite a complicated business; what people should choose to eat and drink on different occasions is not all that simple. A good deal of research has been carried out — and is still being carried out — on food and drink. In the above, I have tried to point out the factors that I consider important for the choice of food and drink. In short, there are many factors involved: needs, edibility, availability, the human senses, philosophical ideas, geographical and mental frontiers, economics, nurture from the very earliest years, ideology, social structure, social classes, gender, utensils and enjoyment. All this gets even more complicated when you realise that the above factors do not necessarily have the same meaning for everyone within a limited geographical area. This chapter began with a review of the words connected with food, and even that gave an idea of how complex the matter is. Food culture also consists of aspects like diet, foodstuffs, culinary art and nutrition. Thinking of the original meaning of the word gastronomy (knowledge of the stomach or a summary of the rules governing the intake of food), I believe that the complexity described above can be summed up as follows: the eating and drinking man is in fact gastronomic man.

Chapter 3

A Gastronomic Mix

All in all, 48 nations received Nobel Prizes during the period 1901–2001, 42 of which were represented in Stockholm and at the banquets that follow the presentation of the prizes. Not all these nations were present at one and the same time, but all the laureates still alive were invited to the jubilee celebrations in 1991 and 2001. This meant that the following nations were present in the Blue Hall during the 90th anniversary in 1991: Australia, Belgium, China, Colombia, France, Great Britain, Italy, Japan, Mexico, the Netherlands, Nigeria, Pakistan, Poland, the Soviet Union, Spain, Sweden, Switzerland and the United States of America. At the centenary celebrations in 2001, laureates from Argentina, Australia, Belgium, Canada, China, Denmark, France, Germany, Great Britain, India, the Republic of Ireland, Japan, Portugal, Russia, St. Lucia, South Africa, Sweden and the United States of America were present at the same time.

Bearing in mind the very wide range of these nations' gastronomic spheres, it was essential for those responsible for the food and drink at the banquets to create menus that suited all tastes within that range. Even if the framework of preferences among the nations listed above was used to create the main menu, there also had to be a readiness and willingness to accommodate the guests who, for religious or other ideological and medical reasons, wished to eat food that complied with their wishes.

This chapter presents a review of a selection of the national food and drink of the laureates' countries listed above. Since there were Jewish laureates during the period 1901–2001, Jewish food, kosher food, and its special rules are presented at the end of the review. The Israel cuisine is also presented to allow a better understanding of the complex Jewish cuisine.

The fact that the nations' food and drink are described at different lengths is not because one nation's food and drink are more interesting or exciting than another's; it is quite simply due to the amount of information available about each nation. I have to a large extent used cookery books when writing this chapter. When looking for information, I have sought books that contain not only recipes but also historical background material. It may seem strange to use cookery books as source material, since their authors themselves make a selection when collecting their material. Nevertheless, I have chosen to look for information in this way although I realise that this may mean that not all that is available is included in this book.

Argentina

Argentina seems to be the only country in South America that can match Chile in wine production. It takes sixth place among the countries that produce most wine in the world and is the largest wine-producing country in South America. The Argentinian wine districts stretch from the province of Salta in the north (close to Bolivia and Paraguay) down to the province of Rio Negro in the south, a distance of some 10,000 km. The largest region for sparkling wines is Rio Grande. San Juan and Mendoza are two other important provinces in which almost three-quarters of the country's vineyards are located. Mendoza has a hot, dry climate and night temperatures are low, which allows the grapes to develop both taste and aroma. By using slightly more unusual grapes, Argentina has taken the lead over the other wine-producing countries in South America. Some of its best red wines are made from Malbec grapes. These give a wine that is deep red in colour and tastes of plums and other black berries. The grapes most used for making white wine are Semillon and Torrontes.

Argentina is famous for its beef cattle; every restaurant has a large grill for grilling various cuts of meat. Beef plays a dominant role in the country's cuisine, which is strongly influenced by Italian and French cooking. Immigrants from these countries brought with them dishes that have been adapted to their new host country. An empanada is a speciality consisting of thin, soft pastry that is filled with beef or pork together with fruit like peaches, pears and the like. Half the pastry is folded over the filling and the edges are rolled or pressed together, making it half-moon shaped. There is also a stew of nuts, squash, peaches, and maize that is boiled slowly and then served in a hollowed-out marrow.

Even though these dishes are characteristic of the country in general, there are also local dishes, because Argentina is divided geographically into four different zones. The most central areas (pampas) have dishes in which pork plays an important part; for example, there are various sausages and roast or grilled meat, and roulades that use slices of meat filled with boiled egg, onions and herbs. These are often served with various kinds of sauces made with garlic, vinegar and different sorts of strong herbs. From the pampas come empanadas that are filled with egg, meat, olives and tomatoes. These are fried golden brown on both sides in oil in a frying pan. There are also stews made of beef or chicken. Other dishes use tongue as the main ingredient; these dishes are often salty. The classic dish from this area, however, is made from beef and various kinds of sausages boiled together that are served with boiled corn on the cob, yuca (cassava), carrots, chickpeas and marrow. A large number of desserts that use milk as the basic ingredient also come from this region. The milk is boiled with sugar and various flavours like vanilla, cinnamon and lemon. When it has boiled down to the required quantity, it is removed from the stove, poured out of the saucepan and allowed to cool, sometimes while being stirred. When it has cooled, it is made into various shapes and eaten. There is also a dessert made of rice boiled in milk that is somewhat like the cold creamed rice called ris à la malta in Sweden. There are also various kinds of sweetmeats in the form of biscuits made of flour and quince.

The northwest parts of Argentina are mostly mountains and the soil there is not very fertile, but there are a number of regional fruits that are used to produce juices. A lot of chicken is eaten here but not much potato, which may seem strange, since the region produces 270 different sorts. The people here are very hospitable. Empanadas are eaten here too, filled with egg, grapes, aja (a strong spice) onion, bouillon, tomato, chopped meat and various fresh fruits. They also eat mock turtle soup, various kinds of beans cooked in white sauce and chicken with a large variety of flavours. In addition, there are many dishes based on dried salted meat from various animals. As for cereals and vegetable, maize, beans, strong spices, garlic, marrow, tomatoes, olives and peppers are the ones most used in cooking. There is also a dish called humita. The corn is taken off the cob and ground and some sort of liquid is added together with spices and salt to make a paste. Various kinds of meat, cheese, sugar, cinnamon, vanilla and so on are then added to the paste. The paste is then mixed well and placed on banana leaves that are folded over. Finally, the packet is tied together with string and boiled. The ones that contain sweet spices are eaten as desserts and the others as main dishes. Potatoes are used mostly in stews. The region's desserts often consist of dried, crystallised fruits and various kinds of pastries with these fruits.

The areas in the northeast are characterised by staple foods like eggs, cheese, fish and different kinds of flour. Some soups here are made with eggs and cheese, and there are delicious stews containing meat, potato, tomato and onion. There are pies too, filled with chicken, egg, onion, tomato, oregano, parsley and pimiento. Fish is an important food and the usual way by far of cooking it is grilling, preferably when the fish is newly caught, but it is also baked with a topping of eggs, carrots and mayonnaise. Boiling and au gratin are other methods used, eaten then with plenty of root vegetables. Thanks to the rivers that run through the region, there is a rich variety of species. Chicken is a favourite food eaten like fish, grilled or roasted, preferably on a spit. Other popular dishes are tortillas made with blood which are similar to thin Swedish blood bread. A great deal of white bread is eaten in this region, in contrast to the dark bread that is usual in other parts of the country. The desserts here are based largely on honey, and the region also has a large number of fruit liqueurs.

Generally speaking, the southernmost regions of Argentina have two different kinds of cuisine, one in rural areas and the other in towns. Country cooking often has heavy meat dishes and various kinds of home-made jams. The town cuisine is more like the French "nouvelle cuisine". Whether you are in the countryside or in a town, there will be a number of dishes based on what the rivers and the sea provide, fresh fish, fresh shellfish and a wide range of herbs like salvia and tarragon and a great variety of salads. Omelettes filled with fresh fish and shellfish are often served in this region, which means that they are seldom the same from day to day. Fish is also used to make pâtés and pies of many different kinds and tastes, but garlic is common. There are also fish and shellfish dishes whose consistency is something between a sauce and a soup. Dishes based on shellfish in particular are also commonly served in large shells and are grilled quickly to give them a golden brown top. Rice is most frequently served with these fish and shellfish dishes, and saffron is quite commonly used to spice them. Trout and salmon are very common on the menu, and mussels and molluscs are often used in fish stews. Fruit is used in both the rural and the urban cuisines; fruits like cherries and strawberries were introduced to the region by German immigrants. For dessert, a tart or a large cake of some kind is popular, as well as various sorts of jellies. Tea is drunk all over Argentina, accompanied by various pastries that might contain aniseed and be baked with corn flour.

Australia

In contrast to countries like France and Germany, Australia has no long tradition of wine-making to take into account or build up on. This means that its wine producers have been willing to experiment in order to produce good wines. The result of this desire to experiment is that Australia has developed wines that have more flavour than those from France, for example. Australian Cabernet Sauvignon wines have the taste of great black wines and the Chardonnay wines have a fullness of tropical fruits with a touch of vanilla. Shiraz wines have a concentrated spiciness.

It is difficult to think of any dish that is typically Australian, which is a result of Australia having been and still being a country of immigrants. This has left its mark on what is eaten and drunk. Nevertheless, there are a few dishes that are typically Australia. One is a soup based on kangaroo tail (similar to oxtail soup) and another a meringue cake called Pavlova, which is a light, sweet dessert oozing with calories. Apart from these dishes, it is barbecues that Australia is known for. These meals can vary a great deal, on the one hand from formal events at which smartly dressed guests sip champagne while waiting for the whole roasted lamb dressed with fresh herbs, on the other hand to people in leisure wear drinking beer round a fire in the company of good friends. The Aboriginals were perhaps the first people in the country to realise the advantages of cooking in clay pots. For a very long time they had wrapped up the animals they had caught, lizards and snakes, in clay and left them for hours in the boiling sun before eating them when they were cooked in their clay packets. A tradition that has survived among elderly women is to preserve all sorts of foodstuffs when they are harvested. They fill their cellar shelves with row after row of jars of preserved food. Some women even take empty jars with them on excursions and camping trips so as to be able to preserve the newly caught fish on the spot. But the younger generation prefer to freeze their food instead of using the time-consuming older methods. It is necessary either to conserve or freeze foodstuffs because more grapefruit, lemons, beans, peppers, tomatoes, cauliflowers, corn on the cob and potatoes are produced in people's gardens than any household can consume within a reasonable time. At Christmas, a traditional English lunch is served (because many people at some time emigrated from England to Australia) And perhaps — although it is in the middle of the Australian summer — roast lamb, goose, small pies and Christmas pudding are on the menu.

In the 19th century, Australia, like many other English colonies, began to produce cheeses like those in England. Cheddar is the most common sort. It is produced on a large scale in the southern parts of the country. Some of the best cheeses come from the Margaret River area. A mild cheddar is stored for three months, a mature cheddar for six to twelve months and a vintage cheddar for up to two years. There are also spiced varieties flavoured with garlic, caraway, bacon and nuts. Even though cheddar holds a unique position, very good versions of the world's most loved cheeses are produced as well, from Emmentaler and Gouda to Limburger and Gorgonzola. These cheeses are mostly made from cow's milk, and Australia has magnificent dairies that are administered by the state.

Canada

Canada's wine speciality is icewine. This is best known in Germany and Austria, where it is called eiswein. Icewine is now produced regularly in Ontario and British Columbia. Most of the wine industries in Canada are very young, only starting in the late 20th century. Small quantities of wine are also produced in Quebec and Nova Scotia. The rest of the country is not suitable for vine cultivation owing to the hard climate. In Ontario, hybrid grapes like Baco Noir, Maréchal Foch, Seyval Blanc and Vidal are grown. The best icewines are made from the last named. Chardonnay, Gewurztraminer and Riesling are the dominant grapes. The red wines are made from Cabernet Franc, Cabernet Sauvignon, Gamay, Merlot and Pinot Noir. The vineyards in British Columbia are located further north than those in Ontario and are centred on the OkanaganValley. This area has a very low rainfall and high summer temperatures, but the day temperatures vary greatly and the nights are cold. All these factors are important for helping the grapes to develop a strong flavour. Riesling is the dominant grape, but Auxerrois, Chardonnay, Pinot Blanc and Gewurztraminer are also used. The red wines are made from Merlot and Pinot Noir.

So much for wines, but Canada is probably better known for its whiskey production. This whiskey is usually light and very often has a vanilla flavour from oak casks and a slight fruitiness from additions of, for example, plum wine. It is made from several types of grain: rye, barley, wheat and maize. The latter is dominant in present-day production and the blended whiskies often have 70–90 per cent of maize spirit which is distilled continuously in great column stills. For the rest of the blend, the so-called flavour whiskey (which accounts for much of the taste), small stills of the column or pot type are used, as in the production of Scotch malt whisky. The larger the amount of flavour whiskey in a blend, the stronger its character. This is often related to its age; blends that are 10–12 years old often have a larger content of flavour whiskey and are sold as expensive premium whiskies. When making all whiskey, the various types of spirits are distilled separately before they are blended. Whether the blending itself takes place before or after maturing varies from one manufacturer to another. The whisky called "blending at birth" has newly distilled spirits that are blended and then matured. Other types are made in the traditional fashion, that is, blending takes place when everything is matured, in more less the same way as Scotch whisky is made. Some producers top off their whiskey with a little sherry, port or plum wine.

Canada does not have a homogenous cuisine; it varies according to the region. The part of Canada that Europeans first came to was Atlantic Canada. Newfoundland is an island surrounded by icebergs and ice-cold water. The food that the first Europeans could make was spartan, based on what could be hunted, cultivated or gathered. Parts of this cuisine still survive in various forms: dishes based on fish (cod's tongue, for example), root vegetables, meat (stews), berry pies and tartlets. These dishes compete with lobster and consommés made from game birds, for example. In principle, Nova Scotia is surrounded by the sea and has therefore always had a strong element of seafood in its cuisine, even today. Other frequent foodstuffs are a wide variety of fresh fruits and vegetables. In Moncton in New Brunswick, recipes have been handed down orally from generation to generation and are consequently very old. This food is typically heavy, using plenty of molasses, syrups, mussels and lobsters. Lobsters were long considered to be poor people's food in this part of Canada. In southern New Brunswick, there is a town called St. Andrews-by-the-Sea which is well known for its wide variety of fresh herbs and its white strawberries. On top of this, it has juicy salmon, oysters, blueberries and much else to choose from. Here they drink apple tea laced with apricot brandy as well as dark tea laced with crème de cassis. Prince Edward Island is Canada's smallest province and has most of the country's best shellfish farms. Blue mussels from the Murray River and oysters from Malpeque are world-famous for their delicate taste. This island also produces a number of berry wines.

Traditionally, Quebec is the most dynamic gastronomic part of Canada, perhaps because of its French heritage, as it was mostly French people from Normandy, Brittany and Poitou who emigrated to this city. It is said that it was their tastes and traditions that created the distinct and delicate food that is now called French Canadian. Pies play a big part in this cuisine, and one of the best-known is made with ham and is only served on Christmas Eve. Various kinds of chicken dishes are also specific for this cuisine. Another animal that plays an important role is the pig; all its parts are used as food. In the province of Ontario, goose is the speciality, preferably filled with stuffing. The parts of Canada that were last colonised are the western prairie provinces. The cooking here is very varied, with Scottish, Ukrainian, Russian and German food side by side, without creating any problems. In British Columbia, which is the country's westernmost province, afternoon tea is still served in the English tradition. This province also has an abundance of fish and different types of berries. In the far north is the province of Yukon, with its virgin areas full of deer, moose, snow geese, hares, seals, salmon and wild berries.

China

Chinese cooking is a paradox; it originated in poverty and misery, and this country has 22 per cent of the world's population who have to be fed from 7 per cent of the earth's cultivable land. No one can agree about how many regional cuisines China has, but most experts distinguish four major regions: Cantonese, with its centre in the south of Guangdong province and Hong Kong; Sichuan, which dominates the cuisine in this western province in the two major cities of Chengdu and Chongqing; Henan, that is, eastern China with Jiangsu, Zhejiang and Shanghai; and Peking (Beijing), which is also called the northern cuisine, whose inspiration is mainly gained from the coastal province of Shandong. A fifth province might be added: the southeastern coastal province of Fujian. All these provinces use various forms of ginger, garlic, spring onions, soya, vinegar, sugar, sesame oil and bean paste, but they combine them in completely different ways. The distinctions between the provinces are based not only on the cooking methods but also on the variations and combinations of these ingredients.

The Cantonese cuisine is characterised by its wide range of dishes, by the freshness of the raw materials, by the well-prepared sauces and by the chefs' willingness to experiment and try out new exotic tastes like lemon, curry and Worcester sauce. These chefs excel in roasted and grilled meat (duck, goose, chicken, pork) and in the small dishes that are taken with tea, as breakfast or at lunch (dim sum). These can be sweet, salt, steamed, baked, boiled or stewed and are served from separate bamboo steamers or plates. Eating these dim sums is the same as drinking tea (yum cha). In special yum cha restaurants, the waiters go round with a trolley or tray, tempting the guests with their delicacies. Dim sum restaurants are important social institutions where the locals go to discuss business, to socialise or just to read the newspaper.

The region or province that is famous for its strongly spiced food is Sichuan. This inland region has extremely fertile soil and a population of more than 100 million. Their love of strong spices is said to be connected with the climate. The rainfall here is so heavy that it is possible to get two harvests of rice in many parts of the region. The strong spices have a tonic effect in the humid, cool climate and also act as preservatives. Chili and Sichuan pepper are the most popular ones, moderated with vinegar, sugar and salt. However, there are dishes from this region that are not strong at all, like the famous tea-smoked, steamed duck, which is smoked over a mixture of tea and camphor leaves.

The core of the Henan province cuisine is in towns like Huainan and Yangzhou. Freshness is the prime criterion for food from this region. The word freshness (xian) means, for example, that a fish should be served within an hour after it has been swimming in the water, so that it still has its natural taste, is soft but is not overcooked. Xian also means that the raw materials' natural taste should not be drowned by the sauce. Many of the dishes from this region are steamed or stewed and take longer and need lower heat to prepare, for example, chicken with chestnuts, pork steamed in lotus leaves and duck with a filling of eight ingredients. This region is best known for its fish and shellfish dishes.

The Peking (Bejing) cuisine is the one that has been most strongly influenced by outside gastronomic ideas. Lamb was introduced by the Mongols (1271–1368) and is the main ingredient in Mongolian stew, a favourite in Peking during the autumn and winter. Various cooking methods for pork were introduced by the Manchus (1644–1912), but the most important influence has come from Shandong province, with roots going back to Confucius (c.550 B.C.). This cuisine includes sea delicacies from the east coast like mussels, octopus (dried or fresh), sea cucumber, molluscs, crabs and shark fins, which are often combined and matched with the flavour of raw leeks and garlic. The famous Peking duck was inspired by culinary districts all over China as well as by the capital itself. The roasting of the duck is taken from the Henan cuisine, while the pancakes, the raw leek and the salty sauce are typical of Shandong. Peking is also well known for its steamed and boiled jiaozi, filled with a mixture of pork, cabbage or leek or with a combination of egg and vegetables.

Breakfast in China may consist of a fried egg wrapped in a pancake, a large, round slice of fried bread, noodles, rice pudding or bean jelly with crisp fried pastry, a slice of cake and a mug of soya-bean milk. Lunch or dinner often consist of noodles from one of the stalls in the food market, if you don't go home and turn your groceries from the market into a fine meal. Every region in China has its own snacks that are usually sold in the street; they are usually cheap and include regional specialities. Examples include Peking's boiled cow's stomach with fresh coriander, fried sausage rich in starch and garlic, fermented bean soup or boiled pork with leek dumplings. Shanghai is known for its steaming bazoi dumplings, a sort of sticky rice made of eight ingredients like fruit, nuts, berries and the like according to the season, and wine dregs rich in yeast. Sichuan has its spicy dandan noodles served with a strong sauce and bean jelly, while the Cantonese dim sum is a cuisine in its own right. The average Chinese who lives in a town eats his main meal in the evening. This meal usually consists of basic food like rice or noodles served with one or two dishes, one of which contains fish or meat, and soup. In the north, more wheat than rice is eaten, often in the form of steaming rolls or noodles that have been fried or boiled in bouillon.

Rice forms an essential part of a Chinese meal, especially in the southern regions. A large number of types of rice are grown and these are supplemented with expensive Thai rice that is sold in all the large cities. In the southern regions, long-grain rice is preferred as it becomes less sticky and has a slight flavour and smells of wood if it is served piping hot. The Chinese have always divided their food into two main groups: boiled rice and the most common sorts of grain (fan); and cooked meat and vegetable dishes (cai). A well-balanced blend of grains and cooked food has always been the ideal for a Chinese meal. A balance between cooling and warming properties must also be observed. The proportions between the different types of grain and the cooked dishes depend on people's financial state and the level of festivity. The more festive the occasion, the less grain and the more cooked dishes. This is a very old tradition that is retained at large banquets, where a symbolic bowl of ordinary boiled rice is served after a large number of delicious dishes.

Rice can be served boiled, fried (after having been boiled) or as noodles made from rice flour. It can also be boiled in plenty of water to make a kind of gruel. This gruel is popular for breakfast or late at night accompanied by other more tasty dishes. Rice is eaten from a bowl that is lifted with one hand to the mouth while the rice is shovelled quickly into the mouth by means of chopsticks. A Chinese meal means a shared meal consisting usually of three or four cai, fan and a soup. All these dishes are shared among the guests, who help themselves. The cai dishes should each consist of one main ingredient, for example, one meat dish, one fish dish and one vegetable dish. Each dish should also complement the others with regard to taste, consistency and smell, so that the total impression satisfies both the eye and the palate. Tea is drunk either before or after the meal but very seldom during the meal. The usual drink at home is beer. Spirits are reserved for special occasions. If alcohol is served with the food, it is Maotai, drunk from small glasses or cups without handles and in one gulp. Thus a Chinese meal is an important social event; special menus are prepared for weddings and birthdays. Important festivals also have their own traditional dishes and trimmings.

Garlic, ginger and spring onions are indispensable in this cuisine, together with plenty of fresh jasmine rice and dried egg noodles. Soya, rice wine, chilli sauce and sesame oil are essential seasonings. Bean paste and bamboo shoots are normal ingredients in everything from quick fries to stews. Sesame paste is mixed in dips and Sichuan pepper gives a subtle warmth to dishes; candied sugar is also used in braised dishes. One should never point with one's chopsticks at a meal, nor stick them into the rice bowl so that they stand up straight or crossed. Neither should chopsticks be used to stir up the common rice bowl in order to find the tastiest bits. You should pick out the bit you want with your eyes and then lift it up with your chopsticks without touching the rest of the food. If you want to enjoy a glass of wine with your food at a formal dinner, you must first toast another guest, whether he replies or not. If someone toasts you but you do not want to drink, it is enough to raise your glass to your lips. It is part of the whole performance that the host plies his guests with more food than they can eat. This means that more food than is needed is brought in. Courtesy requires that the guest of honour is given the tastiest bits, for example cheek meat of fish, which are served with a pair of common chopsticks or the top ends of one's own.

Colombia

Colombia's gastronomy may be described as a mixture of the native and the Spanish cuisine. The food is mostly strongly spiced, but that does not mean that the hot chilli sauce that is always on the table is used all the time. In fact, it is used sparingly. The country's long coastline provides many fish and shellfish dishes. Other important dishes are those made with coconut milk, such as rice cooked in coconut milk with raisins. A variant of this dish is rice fried with coconut milk and raisins. The specialities of the mountain regions differ in part from these rice dishes; examples of dishes from these regions are potatoes with a cheese sauce and stuffed meat.

Fruit juices are popular drinks, but the tap water is not always so good, so it should be boiled before drinking it or using it to make juices. The fruits used include oranges, mandarins, limes, lemons and tomatoes. A very common dish all over the country is salted potatoes. When the potatoes are served, even more salt is sprinkled on the skins. Salads made with fresh vegetables like tomatoes, lettuce and cucumber are less common. Plantains in various forms are more common, but bananas are not so popular. When the plantains have turned yellow and are getting sweet, a cut is made in the skin and they are filled with chocolate and/or cheese and put in a low oven until they are soft.

Apart from the above-mentioned coastal dishes, here, as in Argentina, there is a dish based on corn flour that contains meat and salt. This is wrapped up in a banana leaf, boiled and served piping hot. Fish is served in many forms, prepared in various ways, grilled and in stews. Small fish are cooked whole in stews. Another popular dish is sliced plantain which is lightly fried. When the fat has run off, the slices are flattened out and fried again before serving. As for shellfish, they are eaten in large quantities: mussels, oysters and molluscs. They are also used in a wide range of dishes, just like fish. Fresh fruit like pineapple, coconut, melon, papayas, mango and kiwi are also popular.

Around Santa Fé and Bogota, people eat various ajis, which are very strongly spiced sauces with a number of different bases. Here, too, there are dishes consisting of cooked meats such as beef and goat sausages, roast ham, grilled cutlets and the like. The soups are often made from chicken and various vegetables. The desserts are very sweet: pears, peaches and so on, boiled and preserved in syrup. There are also small, very sweet cakes such as meringues and pastry twists. In nearby districts, they make tasty stews with rice, tomatoes and large pieces of meat, grilled and fried fish, chicken and guinea pigs roasted whole; they also have the same sort of maize packets as described before, but here with a sweet filling and eaten as a dessert. Tongue is often eaten boiled with some kind of sauce. Plantains are common food, prepared in a number of different ways, boiled or baked. When they get a little too ripe and are sweet, they are cut open, filled with cheese and baked. In the Amazon region, they often cook in clay packets. If, for example, you have a fresh bird, you pack it in clay and put it in the embers of a fire until it is cooked. You then make a hole in the clay, pluck all the feathers and serve it as it is. A not unusual dish in this part of the country is fried ants, ants with a large body which are fried in oil.

Denmark

Everyday lunch in Denmark is not a big meal. If you are at work, you usually have a 30-minute lunch break. Most people take their own food to work with them, often rye-bread sandwiches of some kind, but it has become more common to exchange the rye bread for pita bread or bread rolls. Lunch at the weekend or at parties is somewhat different. If you are having guests, it often consists of three dishes, fish (various kinds of pickled and marinated herring, fried fish and shrimps) meat (for example, forcemeat balls, chops with mushrooms and mashed potatoes, liver paste with bacon, or boiled pickled pork with red cabbage) and cheese (all sorts, from cottage cheese to well-matured hard cheeses, with biscuits, radishes, grapes and chocolate twists). Beer and schnapps are the drinks. At festivals like Christmas, Easter and Whitsun, lunch may also consist only of dishes connected with these occasions. Examples of these are cabbage in various forms, brawn, which is eaten at Christmas, and traditional egg dishes at Easter. A Danish speciality is smörrebröd, washed down with beer and schnapps. Smörrebröd consists of open sandwiches with different fillings, often piled up on each other. One of the most popular sorts consists of a fillet of fish with rémoulade sauce and slices of lemon, smoked eel with raw egg and leek, roast pork with red cabbage and a slice of orange. A meal like this is always rounded off with a piece of cheese.

Brunch came to Denmark from the US and to start with was a Sunday event when one chose to sleep in and skip breakfast. Now it is possible to eat brunch at certain restaurants between 11 a.m. and 3 p.m. If you have guests to brunch, it often starts with a Bloody Mary — vodka, tomato juice, lemon juice, Worcester sauce, tabasco, salt and pepper — served with ice and slices of lemon. If you don't want the vodka, that's fine, but then it's called Virgin Mary. After that comes coffee and/or tea and various juices; on more festive occasions, these drinks are replaced by champagne or sparkling white wine. Various kinds of egg dishes are always included in a brunch, eaten with different sorts of bread. Bacon, small sausages, chopped steak and spinach can be served with the eggs. Since brunch came from the US, American pancakes with maple syrup may also be included. Then follows cheese and in the middle of the table there will be a bowl of fresh fruit. But supper is the main meal of the day, usually eaten around 5–7 p.m. On weekdays, this consists of one dish with a salad. Boiled potatoes, sauce, vegetables and meat are usual at weekday suppers, but rice and pasta have begun to replace potatoes due to increasing influences from countries like Italy, Greece and France. If dinner is served instead of supper on festive occasions, it usually consists of three courses, to which are served white wine, red wine and Madeira with the dessert. The first course might be soup or a fish dish; if soup, it will be a consommé with meat or meatballs in it, or asparagus soup. If a clear soup is served, it is often accompanied by a chicken or asparagus tartlet before the main course appears. The fish course could be boiled fresh salmon with mayonnaise and shrimps or fillets of fish cooked in white wine with the same trimmings. If the first course is soup, the main course very often consists of the beef that the soup was made from, accompanied by potatoes, sweet and sour sauce and bread. Other typical main courses are roast veal, roast beef, glazed ham and roast pork. Boiled potatoes and sauce are always served with the main course, accompanied by typical Danish vegetables like cauliflower, peas, and beans. Desserts at these dinners are often based on milk in the form of blancmange or mousse, or they use frozen water as the basis, like sorbets. It is very common that the lights in the room are turned off when the dessert is brought in with candles and music. When receptions are held, they are more formal than cocktail parties, so more dishes are served than at a cocktail party. A reception menu, for example, might contain: fresh oysters, marinated salmon, pike paste with dill and mustard, duck-liver mousse, turkey with lime, coconut and quail eggs, glazed fillet of beef with mushroom sauce, organic Danish cheeses and various kinds of fruit flans. Champagne will be served throughout.

In principle, Denmark imports all the wine it drinks, and 90 per cent of it comes from France (Alsace, Beaujolais, Bordeaux, Bourgogne, Champagne, Côtes-du-Rhone, Jura, Languedoc-Roussillon, Loire, Provence, Savoy), Italy (Val d'Aosta, Piedmont, Lombardy, Trentino/Alto Adige, Friuli, Veneto, Liguria, Tuscany, Emilia-Romagna, Umbria, The Marches, Lazio, Abruzzo, Molise, Campania, Apulia, Basilicata, Calabria, Sicily, Sardinia), Spain (Rioja, Navarre, Penedes, Campo de Borja, Cariñena, Tarragona, Ribiera, Valdeorras, Valdepeñas), Portugal (Algarve, Alentejo, Colares, Buselas, Bairrada, Dão, Douro, Vinho Verde) and Germany (Ahr, Mittelrhein, Rheingau, Rheinhessen, Reinpfalz, Württenberg, Franken, Hessische Bergstrasse, Nahe, Mosel-Saar-Ruwer, Saale/Unstruut, Sachsen). Of the 90 per cent, approximately 65 per cent comes from France. Less than 10 per cent comes from Eastern Europe, Greece, Morocco, Tunisia, Algeria, North and South America, Australia and New Zealand. As for which wine is drunk with which food, the classic rules are white wine with fish, and red wine with meat, but these rules seem to be losing ground and it is not unusual to drink a light red wine with fish. There are, of course, other beverages than wine; coffee is the most popular one. It is drunk at any time and is something of a national drink, so it is not surprising that people are invited to afternoon and evening coffee. Afternoon coffee is served at about 3 o'clock and evening coffee at about 9 o'clock. If you are invited to evening coffee, that means coffee and cakes. Until the middle of the 20th century, coffee and cakes consisted of at least ten different kinds of biscuits and cakes; nowadays, the number and kind of cakes served is much more flexible. Traditionally, evening coffee begins with bread and butter, pastry twists and sponge cake, followed by finer cakes filled with creams, fruit cakes and last of all, fancy biscuits. It used to be very common to round off a big dinner with coffee and cakes like this, but today it is just coffee and biscuits and possibly a kranskaka, a Danish cake made of marzipan and icing and decorated in various ways. If you are invited to morning coffee, you will not get cakes or biscuits but a sandwich or something similar. In the old days, morning coffee was a way for women to get together without their husbands and children. Tea-lovers are not so common as coffee-lovers in Denmark. The people who drink tea instead of coffee have afternoon tea about 3 o'clock, with both lemon and milk. The custom of drinking tea with lemon comes from Russia, while drinking it with milk comes from England. Cake is not served at tea as it is with coffee; instead, toast and marmalade or cheese are on the table. Queen cakes, scones or Danish pastries are also common.

France

In France, food is treated with great respect, and at all levels in society there is an innate feeling for and appreciation of all kinds of food and what is characteristic of them.

The French country cuisine, like most rural cuisines all round the world, consists mainly of the farm's own products: vegetables, eggs, milk, rabbit and pork, and occasionally poultry and game when in season. However, the farmer's wife can buy butcher's products, fish, cheese and fruit all the year round. Her culinary standards are high and she knows that good food is important for good results in the fields. In addition, she knows that food is a Frenchman's everyday pleasure. The recipes she uses have been handed down for generations and everyday food is mostly of the kind that can be prepared early in the day and left to cook while she goes off to work on the farm. But on holidays and at the weekend, more lavish dishes are prepared. The menu will include a pâté and the fresh fruit will be replaced by a fruit flan. When there is a family celebration, the local lady cook is called in (every village has one). It's worth mentioning that the lady cooks who are called in for village weddings do some of the best cooking in France. The plans for feasts like this are made weeks in advance and the family spends more time discussing the menu with the cook than they spend on their clothes.

Provence is a historical French province comprising the departments of Bouches-du-Rhône, Var, Alpes-Maritimes, Vaucluse and Basse-Alpes. The city of Nice, which geographically is in Provence, has a history of its own. In common with Marseilles, it was founded by the Greeks and remained a Greek port until the first century A.D., when the Romans made the province a colony with Nice as its capital. In the centuries that followed, this region was ravaged by barbarians, Germans and Arabs. Other colonisers have occupied the city since then, until 1860, when it became part of France and has been so ever since. As in so many other cases, these events have left their mark on the city's cuisine. Olive oil and wine were brought by the Phoenicians. The preparation of dried cod probably dates back to the Portuguese era, while stockfish was the Norwegians' speciality. Couscous came from the Arabs' former North African colonies and pasta and ravioli from Italy.

The Nice cuisine is easy to distinguish from other French cuisines; here they cook the vegetables in an almost oriental style, using strange mixtures of salt and sour (spinach with raisins, sweet onions with anchovies, carrots with garlic, chicken cooked with figs and so on). Some of the dishes in Nice have to be prepared with local products, which means they cannot be made anywhere else; in other words, there are specialities which only this district has. Poutin is one of these specialities. It is a fairly thick, meaty fish that has been caught along the coast between Antibes and Menton for a hundred years. Each fishing boat may catch only 100 kg of this fish, and only for one month of the year. Another fish in the same category is nonat. It is almost transparent and a little larger than poutin and it can be prepared in various ways: dipped in flour, fried and eaten with lemon juice or just fried as it is. It can also be part of other dishes together with chilli peppers, garlic, egg, cheese and vegetables, or be deep fried in butter or made into soup. Poutargue is the roe of a fish that can only be found in Provence. The roe is gathered every year between 15 July and the end of August when the fish are in Berre Pond near Marseilles and the water is calm. After being cleaned and rinsed, the roe is smoked for four days and then eaten with oil and a little red-wine vinegar. The snails in Nice are small and slightly grey or striped and before being gathered and boiled they feed on fennel and thyme. First they are boiled in herb-spiced stock, then mixed with a strongly spiced sauce and then they are put back in their shells.

Other specialities from this region are lamb's head, cut up, washed and grilled with potato wedges and various herbs. The brain and the cheeks are considered the best parts. Goat's head is roasted in the same way, but the spices consist of parsley and garlic. Calf's head simmered in tomato sauce with a little wine is another of these delicacies. Pig's head is stuffed with onion, thyme, chopped carrots, chopped tongue and ham and spiced with brandy or white wine. The head is first marinated and then boiled for eight hours in white wine. Other specialities are the kidneys, liver, heart and lungs of sheep which are simmered with olives and wild mushrooms, cockscomb stew and sheep testicles. The testicles are sliced and fried and then eaten with lemon and parsley. Anchovies pickled in vinegar are an important ingredient in this region's cuisine. All the local housewives make their own anchovies. Another speciality is fish paste made of fresh sardines. Baby thrushes and larks are stuffed with butter, juniper berries and grapes and wrapped in thin slices of salted pork; they are then boiled in white wine spiced with crushed juniper berries and served with croutons. They can also be grilled on a spit over a fire of pinecones, sprigs of rosemary or vine stems. Towards the end of the grilling, large slices of bread are placed under them to catch the gravy. A bunch of watercress is served with the grilled birds.

It is not only Provence that has regional specialities. Here follows a quick glance at various other French cuisines. Artois is in northern France, on the Belgian border. The people there are beer drinkers and there are not many restaurants. Further to the west, along the German border, is Alsace, famous for its pork delicatessen, its freshwater fish and the geese from Strasbourg. The light Alsace wines like Riesling and Sylvaner compete for popularity with plum, cherry and mirabelle wine. Southwest of this province is Burgundy (considered by many to be the gastronomic heart of France), which produces the country's finest wines like Chablis, Côtes de Nuits, Beaune, Mâcon and Beaujolais. This province also produces an abundance of foodstuffs like Bessem chicken, meat and game, freshwater fish, snails, mushrooms and other edible fungi. We have already presented

Provence, so we move on to Languedoc, which is famous for the dishes prepared from its own raw materials: fish soups at Sète and Montpellier near the coast and cassoulets at Toulouse and Carcassonne further inland. Sheep-breeding is also important here. The sheep's milk is used mainly for cheese-making, in particular Roquefort.

Gascony in the southwest corner of France is scented with the wild plants that give rabbit and snail dishes a special taste. In the south of the province, the vineyards lie close together, supplying the table wines of the country. In the home country of the Basques along the Spanish border, there is a strange mixture of French and Spanish cooking. Peppers are used to a great extent, fried in pig or goose fat. The most famous product from this region is probably Bayonne ham. It is eaten either raw, in thin slices, or cut thicker and fried. Bordeaux, best known for its wines, is in the west of France, producing Sauternes wines and red wines like Médoc. Armagnac brandy is also produced here. Between Aquitaine and Poitou lies Dordogne, where every butcher produces his own tinned goose liver and preserved truffles. From the Loire valley come the light Anjou wines (including Muscadet and Vouvray). This valley is also famous for its freshwater fish — salmon, tench and bream. Along the north coast are Brittany and Normandy, both well known for their fish and shellfish. Normandy also produces milk and cream which are used to make local cheeses like Camembert and Pont l'Évêque. It has plenty of fruit orchards which are used to make cider and calvados. Normandy is also known for its delicatessen.

Wine is almost a synonym for France and all the classic wines are found in the country. They come from 12 districts: Bordeaux, Burgundy, Champagne, the Loire Valley, Alsace, Côtes du Rhône, Jura and Savoy, Provence, the South-West, Languedoc and Roussillon. These districts produce red, white and rosé wines that cover a scale from still to sparkling, very dry to sweet. The Champagne district is the only one that is allowed to produce champagne, but it also produces cognac and Armagnac.

French wines go very well with food and the wine should always complement the food at meals. Certain combinations of food and wine are unforgettable, such as kidney with one of the great burgundies, Dover sole in a rich white-wine sauce with a fine white burgundy or a liqueur soufflé with Chateau d' Yquem. Of the sweet white wines (excluding champagne), those from Sauternes are probably the best known; Sauternes can be noble and very rich or comparatively light, depending on their origin and vintage.

A good quality wine of these types suits dessert mousses, soufflés and tarts very well. A fine Sauternes goes well with goose liver and chicken-liver pâté. Sweet wines also suit oysters. The light, dry white wines are served with oysters, cold and hot shellfish, fried fish, cold meat and egg dishes. Typical wines in this category are Rieslings from Alsace, Muscadet, Sancerre and most of the wines from Pouilly-Fuissé and Chablis. The full-bodied, dry white wines are served with fish, poultry and veal in cream sauce. A white burgundy also goes well with goose liver. Rosé wines can be served with all types of food but are usually kept for cold dishes, pâtés, eggs, and pork. Light red wines that are typical of the Bordeaux wines from Médoc and Graves are served with roast chicken, turkey, veal, lamb, beef, ham, liver, quail, pheasant and goose liver. Table wines in this category suit simple dishes like meat and lamb stews, pot steaks, bouillabaisse, joints of meat and pâtés. Full-bodied red wines like all the great burgundies and Rhône wines are served with duck, goose, kidney, well-hung game and meat marinated in red wine. These wines are always suitable with strongly-flavoured food. Champagne is served as an aperitif or at the end of an evening; it can also be served throughout a meal if it is a dry champagne. A medium-dry champagne is served as an aperitif or with shellfish, goose liver, nuts and dried fruit. A medium-sweet or sweet champagne is served with desserts and pastries.

Another food that France is very well known for is cheese, and of course Frenchmen drink wine with their cheeses. The right combination of cheese and wine is sometimes called the world's most successful marriage. The real connoisseurs bring out their finest Bordeaux wines or burgundies when they eat cheese. In general, light red wines like Beaujolais that have a bouquet and a grapey flavour are recommended for white cheeses like Brie and Camembert, while blue-vein cheese like Roquefort or Bresse Bleu need more full-bodied wines or port. Cheeses of the Gruyére type go well with a Gewürtztraminer, which also suits goat cheeses, as does a white Burgundy. Red rind cheeses like Livarot, Münster and Maroilles require a full-bodied red wine with a bouquet like St Émilion or Graves. Goat cheeses go best with a very dry white wine tasting of grapes or with a wine from the same district as the cheese. Semi-soft cheeses like St Paulin and Tomme require a light, delicate white wine or a rosé wine tasting of grapes such as a white Burgundy or a white Anjou. Green cheeses go well with a light white wine and a rosé table wine. Dry, hard cheeses like Comté are best with a dry white table wine or a full-bodied red wine with a bouquet. Processed cheeses require light, dry wines of all three categories.

The French not only drink wine, they frequently use it in cooking as well. It is important when using wine in cooking that it is the aroma that remains in the food, not the alcohol, which is vaporised. That is why wine as well as other spirits that are used in cooking should have a good taste. If the wine is too fruity or acidic or tastes unpleasant in some way, this will be reflected in the finished dish. That is why the French (and, hopefully, others too) use strong, dry and never acidic or fruity white wine in their cooking. A white Mâcon made from Pinot Blanc or Chardonnay grapes is a good choice. When using red wine, it should be young, full-bodied and sweet. A Mâcon, a simple Burgundy or a fairly full-bodied Bordeaux like St. Émilion is a good choice as they have the desired qualities. Other spirits like rum and liqueurs are drunk, for example those with an orange base such as Cointreau, Grand Marnier and Curaçao. Madeira and Port are used to give a relish to sauces. Brandy is the most indispensable spirit in French cooking, used in everything from desserts to sauces, bouillons and flambé dishes.

Germany

In Northern Germany, the reigning beer used to be Berliner Weisse, a wheat beer that has undergone a second fermentation with lactobacillus so that it gets such a sour taste that people often give it a Schuss, that is, they add red raspberry syrup or green sweet woodruff juice to it. This beer is not spiced, but thanks to the yeast's reaction with the wheat it gets aromas reminiscent of apple, plum, banana and clove. Wheat beer often retains its yeast sediment, which gives it a certain cloudiness and makes it tastier. This is pointed out by the word Hefe (yeast) on the labels, while the word Kristall is on the labels of bottles that contain filtered beer. Wheat beer is a great drink with meals because it is less bitter, as the brewers do not put as much hops in it as in beer made from barley malt. Wheat beer has sweetness and tartness or wine-like qualities. Wheat malt also gives more flavourings than barley malt. Beer is not the only alcoholic drink that Germany produces; it also produces wines that are sometimes used to make a wine cup with pineapple. German wines are characteristically light, delicate and full of aroma and they have a low alcohol content. Mosel wines are unique and even Riesling wines have a special status on the country's wine map. Germany produces mainly white wines and an absolute speciality is Eiswein (icewine).

German gastronomy has been subject to outside influence all through history, from the Romans to Napoleon, and it is characterised nowadays by contrasts. The Germans themselves mix sweet and sour, and hot and cold in their cooking (for example, hot fruit sauce poured over cold ice-cream) as well as colouring their food with strong colours. There are many dishes in which meat and fish are combined with fruit, vinegar and sugar. Germany has retained the old European custom of beginning a meal with soup, often served as a main course together with sour bread. In the north, these soups are heavy, made of dried beans or peas, meat or sausages and contain potatoes or dumplings. Pork (for example, minced pork steaks and cumin schnitzel with onion sauce) and game are also common. Along the coast in the north, herring is fished in large quantities and is popular in fresh, salted or pickled form. In the rivers and lakes, it is mostly pike and pikeperch that are caught. Germany is also well known for its locally produced delicatessen in the form of many different types of sausages, pâtés and pastes.

India

India is a vast country in area, divided into many states. These states stretch from snow-capped mountains in Kashmir to the southern tip of luxuriant Kerala, from the dry, barren deserts of Rajasthan in the west to the distant tribal regions of Assam on the Burmese border. In the cold, northern region and on the Ganges plain in central and eastern India, rice and wheat are the staple foods. In the deserts of Rajasthan and Gujarat it is mainly millet and maize that are grown. Rice is also the staple food of eastern India. Along the coast of Kerala, fish and meat are the most important food. The heart of Indian cooking — spices — are grown on a large scale in most regions. Cardamom, cloves and pepper are cultivated mainly in the south, while Rajasthan, Kashmir and Gujarat are known for their turmeric and chilli. Kashmir also produces the flower that is used to make the world's most expensive spice — saffron. The most important spices in the Indian cuisine are cardamom, caraway, nutmeg, cinnamon, fennel seed and fenugreek. Other spices are chilli fruits, fresh or dried, which are sliced or ground and mixed in various spice pastes with garlic, onion and ginger. Lentils, beans and basmati rice are eaten with the numerous curries. Saffron and turmeric give flavour and colour to the various dishes. Yoghurt is used both in main dishes and in desserts and is also the main ingredient of a popular drink called lassi.

From a culinary point of view, the term spices in India comprises dried seed, berries, bark, roots, flowers, leaves and fruits. These may be dried or fresh, bought as pods or seed, be roasted, ground or soaked in oil, which brings out the taste. Some of these spices are used whole, while others are always ground; some are used with meat as they would be too dominant if they were used with more delicate ingredients like fish or vegetables. Masala is a word meaning spice mixture, the best known of which is garam masala, a fragrant mixture of saffron, nutmeg and mace. These spices are then mixed with the fresh roots and leaves of ginger, garlic, turmeric, mint and chilli. The medical properties of spices are always considered when cooking food, as well as their effect on the vegetables or lentils used and their inherent qualities.

Simplicity reigns in an Indian kitchen, where there is an oven (often coal-fired), a few utensils such as a wok-like bowl, pans with straight sides and a few frying pans. In general, the menu depends on factors like the climate, nutritional balance and religion. The food is often very simple; rice and bread are the most important components. Each region and individual household then contributes its own special methods for cooking vegetables, meat and fish, and its own recipes for pickles and chutney. Every meal is served with rice or bread or both. In the cooler, northern parts of the country bread is most usual, whereas rice is most common in the south. The food is often served on a banana leaf. Washing your hands before a meal is a very important ritual since food is usually eaten with your hands while you squat on the kitchen floor. The family eat together except for the mother or the wife, who serve the food. In middle-class families, it is the maid who serves. Those who are eating sit in a long row, and the person serving fills up the dishes all the time as the food is eaten.

All the food served is placed on banana leaves at the same time, but there is a given order in which the dishes are eaten. The first mouthfuls of rice are eaten with ghee or chutney or some similar spicy paste. Dhal, a lentil or pulse dish, is served with a range of boiled vegetables, flavoured with many kinds of spices and trimmings. Poppadums and accessories are served all the time, as are rice and dhal. Guests at a meal are always given the best bits of fish or meat. Unleavened bread is most common in the north, where it is eaten with dhal and vegetables. The finale of a meal often consists of a milk-based dessert, but in the south is it more usually rice with curdled milk, which is thought to calm down the stomach after a spicy meal. This is followed by something typically Indian: a betel leaf with flavouring, or pan. This leaf is chewed with a piece of betel nut, a few drops of lime juice and a knob of paste made from the extract of a tree. This quid has several functions: it symbolises hospitality and moral and legal commitments, it promotes digestion and it forms a suitable meditative end to a meal characterised by generosity.

In Kashmir and the Himalayas, the food has a subtle blend of fragrant herbs, combined with richness and nutritiousness. One of the more popular dishes is yoghurt-marinated lamb that has simmered in milk spiced with nutmeg; rich meat curry dishes are also popular. In the Ganges region, rice is served followed by vegetables fried with spices, dhal, unleavened bread, yoghurt and a dessert; chutney and pickles are common. Bengal is the only part of India where each dish is served separately, a tradition based on ancient ideas about how the ingredients affect digestion. Fish plays an important part, as do rice, dhal and chutney. In the south, rice appears in many forms as it is a staple food: steamed, puffed, and turned into paper-thin pancakes, dosai. This rice is served with a wide range of chutneys, vegetables and light soups. In Karnataka, a state in the middle of the southern part of India, vegetables are the main food, eaten with dosai, idli (steamed rice cakes) and steamed rice. Cucumbers and the like and aubergines are popular, often served with richly spiced appetisers between courses. Vinegar and mangosteen fruit are used in the Goa region, which is also known for its fiery chillies.

India produces very little wine although it grows plenty of grapes— only a little of which is used in wine production. The best-known and most popular wine is Omar Khayyam, a sparkling wine fermented in the bottle.

Ireland

Irish baking comprises both bread, including scones, and cakes. Whether bread, buns or cakes, they are served with butter, jam and whipped cream. Irish cooking is characterised by its simplicity; home cooking is based on the best that agriculture can produce, for example nourishing soups made from cabbage, potatoes, leeks, carrots and onions with meat added in the form of mutton or bacon. In many Irish households, potatoes are the basis of the cooking. These are combined with other favourites such as white cabbage and leek and enriched with butter. One example is Irish potato puffs made from old, mashed potato with Cheddar cheese, yellow onions, various spices and egg added. This is all mixed together and shaped into balls that are deep-fried brown in hot oil. Another popular vegetable dish is leek in cream sauce. The leek is cut up into four parts which are tied together in bunches and boiled gently in water that has been flavoured with lemon peel and bay leaves. These boiled bunches of leek are taken out of the saucepan and drained off before being put on a serving dish and covered with cream sauce. Golden brown, butter-fried breadcrumbs are then sprinkled on top. A traditional vegetable dish is colcannon, based on boiled potatoes which are mashed with a fork until they are dry and floury; lightly boiled shredded white cabbage is added after it has been fried for one minute with shallots. The dish is then topped with chopped parsley.

As for fish and shellfish, the Irish have a famous shellfish dish consisting of fried Dublin prawns; another genuine classic Irish dish is fish au gratin based on fillets of white fish. Fish is also popular baked or grilled. Turning to meat dishes, Irish poultry, beef and lamb are of very high quality. Their main ingredients are used with great imagination in both rural and urban areas. A very well-known evening meal is Irish stew, made with whole mutton or lamb cutlets, potatoes, leek, bacon and carrots. Another stew is based on beef with Guinness added. A favourite dish is a garlic-stuffed roll of lamb boiled in meat bouillon with a flavouring of lemon, bay leaf, onion and carrot which is served with a caper sauce. Smoke-cured pork with whiskey sauce, rabbit with mustard sauce, chicken pie and roast chicken are other popular dishes.

The classic Irish desserts are filled with goodies like cream, fruit and raisins and whiskey, aromatic liqueurs and Guinness are often added. One example of this kind of dessert is Irish Mist Cream (which should be made the day before it is served). This consists of milk, castor sugar, egg yolks and gelatine that are whipped up in a water bath. Irish Mist whiskey liqueur is then poured into the mixture, followed by hard-whipped egg whites and whipped cream. On top of that come more whipped cream and chopped nuts or chocolate. Somewhat heavier desserts are Guinness pudding with whiskey sauce and apples in potato pie.

The famous Guinness stout from Ireland was first brewed in 1759. It is brewed from water, malt, hops and yeast. Irish Mist is a whiskey liqueur that is said to date back 1,300 years in time; it has an alcohol content of 36 per cent by volume. Irish whiskey is made from barley malt and then matured in barrels for seven years.

Israel

The Books of Moses describe how the Israelites, during their 40-year-long journey through the desert, missed the delicacies that Egypt had, like zucchini, water melon, onion and garlic. There are lentil and bean dishes from this time that are still prepared in Israel. Israel is a fusion of many nationalities, which means that its cuisine bears influences from several parts of the world and from the three world religions that exist in the country: Judaism, Christianity and Islam. Probably the best known of all meals took place here: the Last Supper that Jesus ate with his disciples before he was betrayed, arrested and crucified. Although this meal is widely known and has often been portrayed in art, no one knows what was served at it. However, there is no reason to believe that it differed from other Jewish meals at that time, either in the dishes served or in the logistics. The table was probably already laid with all the meal's dishes, together with the glasses for wine, when Jesus and his disciples sat down to eat. The guests brought their own knives to cut the meat with, but they probably ate with their fingers. As this results in sticky fingers, there were most probably servants standing by with bowls of water so that the guests could wash their hands during the meal. The meal most likely began with a simple vegetable soup (traditionally offered even today). What the next course was would depend entirely on the financial state of the host, but as Jesus and his disciples were noble guests, it is very probable that it consisted of the best the house could offer: roast lamb. Traditionally, no dessert would be served, but at festive dinners the meal would end with the guests eating the fruit and nuts that had been placed on the table for decoration. The chances are very great that this meal ended in this way. Over and above the dishes already mentioned, there was also probably unleavened bread, wine, salt, water and bowls of bitter herbs like rosemary, coriander and thyme on the table.

Judaism has very strict rules concerning what foods may be mixed or not mixed. Christianity, however, has no such limitations on what may be eaten or what foods may be mixed with each other. The Christians in Israel follow the rhythm of the year along with the church festivals to a considerably greater extent than in northern Europe. Like Judaism, Islam forbids the eating of pigmeat, nor is it allowed to drink alcohol in any form. This ban may even apply to guests of another religion. The Koran also forbids the eating of animals that have died a natural death, food that has been dedicated to anyone else but God, animals that have been strangled, beaten to death, cut open, fallen down, wild animals that have been killed on a pilgrimage, and animals that have been eaten by beasts of prey, in other words, every animal that has not been slaughtered at the right time. Even the permitted food is regulated and includes the animals that man's hunting animals have been trained to catch, and sea fish, milk, grapefruit, pomegranates, dates, figs and honey.

Throughout history, the country has experienced much immigration, resulting in a mixed population. Its gastronomy has been influenced by many countries, including those in the Middle East. Syrian influence has resulted in dishes like olives filled with anchovies and garlic soup. Iran has contributed dolmas, chicken with chickpeas, date and orange salad and spicy fish soup. There are Egyptian influences in dishes like bean cake and pecan soup. The Maghreb region has contributed dishes like honey cream and fish with zucchini sauce (Tunisia), spiced aubergine, glazed duck and spicy peaches (Algeria), spicy sausages and marinated chickpeas (Libya), and deep-fried olives with mushrooms and chicken with almond (Morocco). From the countries round the Mediterranean, Turkey has left its mark in dishes like egg with yoghurt, arrack cakes and rice with nuts and Greece has contributed asparagus with eggs and lime sauce, bean soup and honey cake with nuts. Influences from Central and Eastern Europe are found in dishes like cold asparagus with walnuts, tongue in wine sauce and cucumber salad (Hungary), duck with radishes, potato pancakes and rice pudding with fruit (Poland), marinated mushrooms, spinach soup and fish in a mustard sauce (Romania) and finally, cabbage soup and cheese straws (Russia). Besides these influences, Israel's gastronomy has also been influenced by Bedouin dishes like lamb on vine leaves and lamb stew. There are also influences from Ethiopia in the form of spiced chicken wings, lamb soup, green beans with peanuts, chicken in a walnut sauce, apricot-stuffed turkey and red beans with a plum sauce; from Georgia comes cheese bread; and finally lamb meatballs, unleavened bread and roast meat with a hot sauce come from Yemen.

Meze is served and eaten in Israel just as it is in Egypt. Examples of these fresh vegetables to whet the appetite are turkey salad with melon, salat caramel (a salad made with boiled eggs, avocados and lettuce) white cabbage salad and avocados with a piquant filling. Examples of meze made with boiled and fried vegetables are marinated aubergine, onion dolmas, falafel (made with chickpeas), Jerusalem artichokes (a thorny plant, grey-green in colour), beetroot salad, carrot salad and fried slices of aubergine. Other creamy salads served as meze are aubergine and avocado salad, sardine salad and vegetarian pâté. Pies and dishes au gratin are also popular, for example, aubergine and cheese pies and marrow au gratin.

Soups are an appreciated dish and just as in Egypt they are preferably served with tasty, wholemeal bread. Examples of such soups are vegetable soup, soup with filled dumplings, celery soup, cold avocado soup and cold tomato soup. This last soup is called gazpacho and probably comes from Spain, but in the Israeli version it is tomatoes that are the main ingredients, in contrast to the Spanish version. In the category sauces, dips and spice mixtures, there is avocado dressing to go with salads, yoghurt, tehina (sesame paste) and avocado dip.

Fish is usually festive food that is served at the beginning of holidays and festivals in the form of, for example, boiled fishcakes; the way these are cooked varies from one chef to another. But fish like tuna au gratin is also served for lunch and light evening meals. Fried fishcakes, fried fish with spring onions, grilled fish with savoury butter and pickled herring are other popular fish dishes. Lamb and mutton are as common in Israel as in Egypt, mainly due to the fact that neither Judaism nor Islam allow pigmeat. Besides vine-leaf dolmas, there are cabbage-leaf dolmas stuffed with minced lamb or mutton, and this minced meat is also mixed with minced beef and served as meatloaf spiced with cinnamon, cloves and allspice. Meat is also served in the form of stews and especially for Jews in various kinds of Sabbath stews. Grilled turkey breast is popular, while chicken is often eaten roasted with oranges. Chicken liver is either made into a pâté or eaten chopped and fried.

Sweets and cakes are often made from sesame seed and they should be sweet, but sesame seed is not the only ingredient used. The traditional Israeli cake for holidays and festivals is honey cake. One of the most popular cakes in the country is cheesecake. Another ingredient used in baking cakes is dates, both fresh and dried. Sweets made from oranges are extremely popular.

Since Israel has so many immigrants and different religions, a wide variety of dishes has to be served at the various festivals. These festivals include the Jewish New Year, when big parties are traditionally held. The food at New Year might include bean soup, stuffed breast of veal and baked apples filled with raisins and nuts. At the Christian Christmas celebrations, the meal might consist of artichoke pie, spinach soup with sour cream, trout with herbs, chicken with mushrooms and apple cream. At Easter, a meal consisting of vegetable soup, unleavened bread and roast lamb is a likely choice. The Muslim New Year is celebrated with prayers and a meal consisting of spiced olives, aubergine, lentil soup, chicken, pepper salad and desserts made from pomegranates and figs.

Israel also produces a good deal of wine, which has developed from being cheap and sweet, bringing tears to the eyes and burning the throat, to wines that have put Israel on the world's wine map. Certain of these red and white wines compete at present with those from, say, California and Australia. Wine consumption is around 6–6½ litres per person per annum.

Italy

The earliest known printed cookery book in the world comes from Italy. Its author was called Platina and the title is *De Honesta Voluptate et Valetudine*. This title has been translated into English as "On Right Pleasure and Good Health". Platina was the Vatican's librarian and his book was first published in about 1470, the second edition coming in 1472 and the third in 1480. It is said that the cradle of the Western European cuisine is in Italy. This country's gastronomy has its roots in Roman, Greek and Arabian cultures, and with its rich supply of vegetables and fruit, Italian food is undoubtedly more varied than anywhere else on the continent. In the 16th century, the districts round Florence were considered to be Europe's culinary centre. When Catherine of Medici moved to France to marry the future king of that country, she took her own chefs with her so as to teach the French how to cook food.

Until 1861, Italy comprised a collection of independent states, which had their own laws, customs and traditions, and this has left its mark on cooking, among other things, right up to the present day. The differences between northern and southern Italy are particularly noticeable. In the north, for example, the pasta is flat, freshly made from eggs, and the fat used in cooking is usually butter. In the south, on the other hand, round pasta is predominant and butter is replaced by olive oil. The flavours in cooking are stronger and more evident in the south than in the north. Some of the well-known pasta dishes from the north are ravioli and minestrone. In the rice-growing region just above Venice, arborio rice is cultivated, which is a special variety that absorbs liquids very easily. This is the rice that forms the basis of dishes like risotto. From Parma in the north come Parma ham and Parmesan cheese. Parmesan tastes best after being matured and dried for two years. In most regions, fish is a dominant feature of the cuisine. Venice is especially well known for its octopus, shrimps, mussels and its red and grey mullet. In the north, sea fish are complemented by splendid fish from the lakes in Lombardy — principally eel. Around Sicily and the south coast of Italy tuna, sardine, swordfish and shellfish are typical catches. The characteristic aroma of food from the south comes from an abundance of locally cultivated tomatoes, garlic, herbs and anchovies. Naples, which is considered to be the culinary centre of the south, claims to be the home of the pizza and of ice-cream. Italy is one of the major wine-producing countries in the world and almost every district contributes to the vast exports of wine.

Here follows a list of some of the most important ingredients in Italian cooking (omitting the local variants). Amaretti are a kind of hard, crisp biscuits that are made from ground almond, sugar, whites of eggs and crushed apricot kernels — which give the biscuits a slightly bitter taste. These biscuits are eaten as they are or are ground up and used in desserts. Bread crumbs are made from white bread dried in a warm oven. Bouillons made by boiling veal, beef and chicken are used in all kinds of

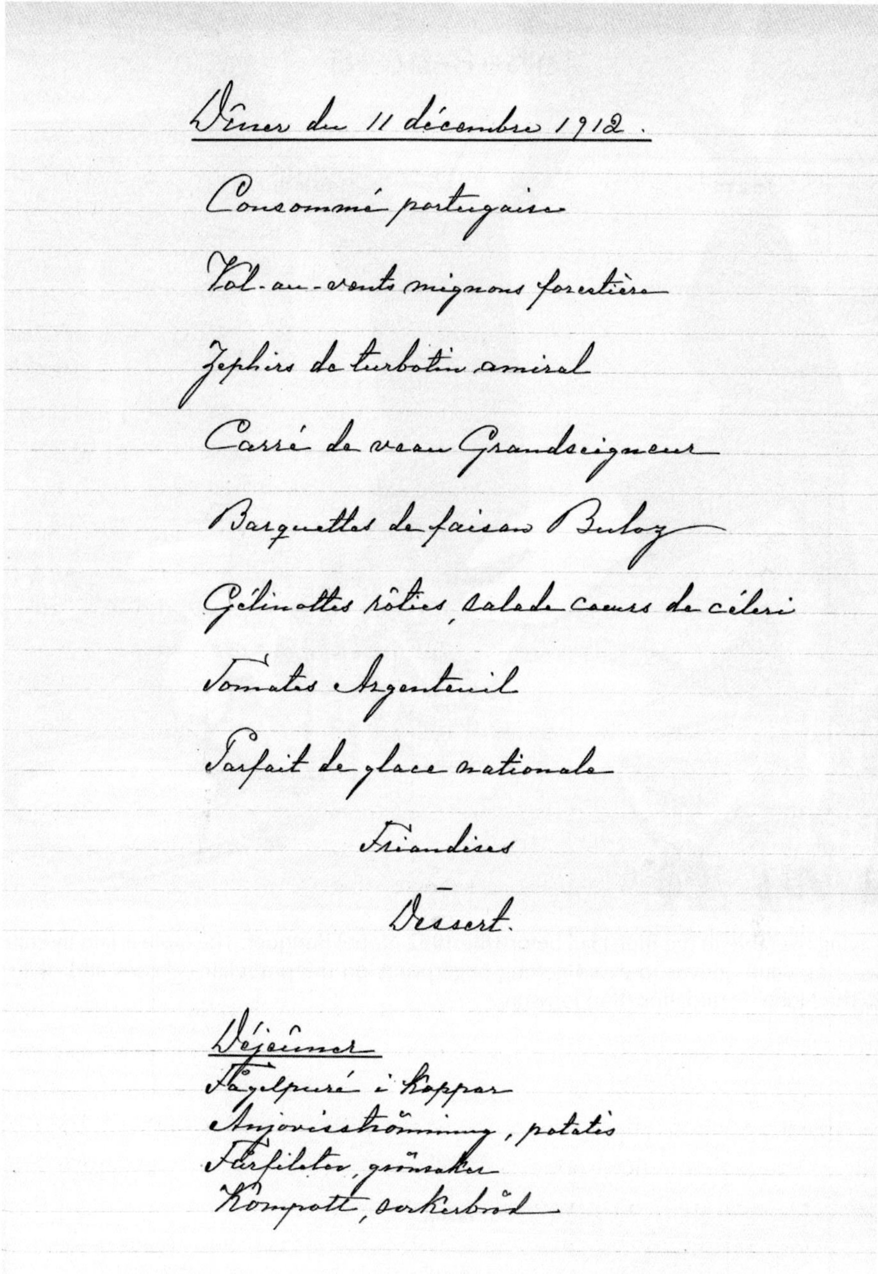
Dîner du 11 décembre 1912.

Consommé portugaise

Vol-au-vents mignons forestière

Zéphirs de turbotin amiral

Carré de veau Grandseigneur

Barquettes de faisan Buloz

Gélinottes rôties, salade cœurs de céleri

Tomates Argenteuil

Parfait de glace nationale

Friandises

—

Dessert.

Déjeuner

Fågelpuré i koppar

Anjovisströmming, potatis

Fårfiléer, grönsaker

Kompott, sockerbröd

The handwritten menu from the banquet at the Royal Palace in 1912. © Slottsarkivet, Stockholm

Laying the table in the Blue Hall before the 1992 Nobel Banquet. The table is laid by staff wearing white gloves to avoid getting fingerprints on the porcelain, glasses and silver. © The Nobel Foundation/Boo Jonsson

The table of honour in the Blue Hall before the 90th anniversary of the Nobel Banquet using the specially designed service, tablecloths and napkins. © The Nobel Foundation/ Boo Jonsson

A photograph showing the precision with which the tables are laid. The covers stand in straight lines before the guests arrive in 2001. © The Nobel Foundation/Boo Jonsson

The royal stewards Helene Bladh and Eva Ohlsson in their beautiful uniforms in 2000. © Kungl.Husgerådskammaren. Photographer: Alexis Daflos

Part of the table of honour on 11 December, 2000. Note that the way the napkins of the King and Queen are folded differs from the way the other guests' napkins are folded. Her Majesty the Queen's cover is in the foreground. © Kungl.Husgerådskammaren. Photographer: Alexis Daflos

A cooked lobster awaits being placed on the plate. © Photographer: Torbjörn Boström

dishes. When Italians use chocolate, they choose a good quality and the same is true of cacoa. Veal is very popular and on average people eat 7 kg per person a year. A distinction is made between prime veal from young calves fed on milk and veal from calves fed on grass. The meat from a milk calf is light pink in colour and tender, and the fat is creamy white. When capers are used, the large, smooth variety are chosen and not the small ones preserved in vinegar. There are many types of sausage, the most popular of which are Salami (in various forms), Luganega (a slim, fresh sausage), Mortadella and Zampone (a pressed sausage). Both green and black olives are cultivated in Italy and are used in its cuisines. In Sicily, they grow both unusually large and small, oval olives that are stoned, soaked in salt water and flavoured with oregano, chilli, fennel and garlic. Olive oil is the only oil that is used for cooking.

Cheese is used in all cooking. Grated, it is sprinkled over pasta and soups, it is mixed in pasta, meat and vegetables and it is served with fruit and wine. Fontina Valdostan is a very characteristic cheese from northern Italy. It tastes of truffles, is brownish-yellow on the outside and has small holes. When it has been matured for four months it is at its best. The best mozzarella cheese is made from buffalo milk. The real Parmesan cheese is called Parmigiano-Reggiano and is made in the districts of Parma, Reggio, Modena, Mantua and Bologna. It has a granular consistency and is damp, salty and yellow in colour and is made by hand according to certain well-defined rules. Pecorino Romano is a hard cheese made from sheep's milk which has a characteristic taste. It exists in several variants: mild, strong, soft and hard. Ricotta is a soft whey cheese, mild and unsalted which is often used as a filling in pasta, but also as a dessert sprinkled with sugar, finely ground coffee, cinnamon and grated chocolate or with a little Sambuca. Pasta is made with eggs and finely-ground durum wheat and exists in many forms.

Flat-leafed parsley is preferred to the curly-leafed variety since it has a stronger flavour. Other common spices and herbs are: red pepper, garlic (fresh), basil (indispensable), bay leaf, marjoram, oregano, rosemary, sage, saffron, vinegar (made from wine), tomatoes, tomato purée and pine kernels. Italian bacon is not smoked but treated with salt and spices; it has a mild taste and is not as fat as ordinary bacon. Truffles are popular and the most sought after is the white kind, which is sold in October; it is very expensive and rare.

Here are some examples of dinners that follow the seasons:

Spring
Pasta soufflé, Veal cutlets in bouillon, Asparagus with herb sauce, Pie with a chocolate filling (Family dinner)

Salami with hard-boiled eggs, beans and sheep's cheese, Easter pizza with ham and mature cheese, Bone-free dill lamb in an egg and lemon sauce, Potatoes with rosemary, Chicory salad with mustard cabbage in an anchovy sauce, Lemon sorbet (Easter dinner)

Rice salad with smoked salmon, Roast veal with truffle sauce, Peas with ham, Vanilla ice-cream with orange sauce (Supper)

Summer
Grandma's stuffed peppers, Cold Wienerschnitzel à la Milano, Beetroot and onion salad, Zabaglione (Cold supper)

Mushroom, Parmesan and celery salad, Piquant breast of chicken in crispy bread, Fried tomatoes with rosemary, Amaretti pudding (Lunch)

Octopus salad, Tomatoes stuffed with rice, Ham with a tuna sauce, Deep-fried peppers, Marinated beef, Marinated zucchini, Mixed Italian salad, Fruit sorbet with stripes, Almond cake (Cold summer dinner)

Autumn
Tagliatelle with bay leaves, Scaloppini with lemon, Spinach with raisins and pine kernels (a weekday dinner)

Pheasant pâté, Loin of pork with blueberry sauce, Braised artichokes, Shallots with rum (Autumn buffet)

Risotto à la Milano, Veal roulades stuffed with artichoke, Braised fennel, Fruit cake (Autumn dinner)

Winter
Soup with pasta and beans, Ham stew with fennel seed and small onions, Lemon pie (Informal winter dinner)

Ravioli with spinach, ricotta, butter and sage, Turkey with hot fruit, Green salad, Soft ice-cream with marrons glacés — a Christmas dessert (Christmas dinner)

Irmana's game croutons, Wild boar in a sweet and sour sauce, Green salad with celery and fennel, Honey caramels with walnuts and bay leaves (Country lunch)

Italy produces about 1,700,000,000 gallons (77 million hectolitres) of wine a year. With this volume, it leads the wine-producing league, beating countries like France, Russia and Spain. In fact, it accounts for one-quarter of the world's wine production. On average, its inhabitants drink 82 litres of wine per capita per annum. Plinius (a writer in the Roman era) describes 91 different varieties of grapes; today, Italy produces 200 different kinds of wine. Thanks to a wide range of soils and the climate, it has been possible to grow a large number of types of grapes, each type adjusting to the district it grows in. Certain types have been cultivated in the same district for thousands of years. One example is Lambrusco, which has been grown for wine production in the southern regions of Italy since the days of the Etruscans and the ancient Greeks.

Wines from Tuscany (Chianti) are made from Sangiove, Canaiolo, Trebbiano, Malvasia and Colorino grapes. In the Piedmont region, the Freisa grape is the most usual type as it prefers altitude, dry soil and an exposed site. Dolcetto wines are fruity with a sweet taste; these grapes are picked four weeks before Nebbiolo grapes. These wines should be drunk young and like Barbera wines are table wines. Wines from Nebbiolo grapes are for festive occasions. In the Lombardy region and especially in Valtelina, Nebbiolo grapes are grown everywhere and are known there as Chiavennasca. Lugana wines are made from Trebbiano and Vernaccia grapes. Wines from grapes like Riesling, Pinot Noir, Bonarda and Barbera are also produced there. Wines from the Veneta region contain six types of grapes: Rondinella, Molinara, Rossignola, Negrara, Corvinon and Pelara. The Giulia, Venecia and Friuli regions are best known for their red wines made from Cabernet Sauvignon, Merlot and Refosco. This last grape is a local variant that is found in France under the name Mondeuse de Savoie. The white wines are made from grapes like Raminer, Riesling, Sauvignon, Chardonnay, Pinot Gris and Pinot Noir. The grapes that are characteristic of the white wines from Emilia Romagna are Sauvignon Blanc, Pinot Gris and Pinot Noir, Tokay and Riesling. The red wines are made from Merlot, Cabernet Blanc and Pinot Noir. From the Umbria region come wines made from Verdicchio grapes, a variety that has been grown in the area since the 14th century. This is a sparkling type of wine.

In the Ambruzzi region, the white wines are made mainly from Trebbiano d'Abruzzo and the red wines from Montepulciano d'Abruzzo. The rosé wines produced in this region are made from Cerasuolo d'Abruzzo. The above-mentioned types of grapes are only a small selection to demonstrate the rich variety of viticulture in Italy.

Japan

A Japanese meal can be divided into three parts: a beginning, a middle and an end. The beginning consists of appetisers, clear soups and raw fish; the middle part is usually a number of fish, shellfish, meat, poultry and vegetable dishes, grilled, steamed, boiled, deep-fried or pickled (a vinegary salad); and the last part consists of rice, miso soup and pickles, green tea and fruit. In order to make sure that there is variety in the middle part of the meal, it is important that each type of cooking is used only once. If, for example, the fish has been fried, the meat will be boiled and the vegetables steamed. An alternative is that the different dishes in the middle part are replaced by a stew, nabe, in which vegetables, fish, meat, bean paste and noodles are combined in one dish. The choice of dishes depends on availability, the season and how much time can be spent on preparing the meal. The Japanese do not classify food according to the raw materials used but according to the method of preparation. Thus it is categorised as grilled, steamed, boiled, deep-fried and pickled in vinegar. In Japanese homes and at restaurants, all the dishes are placed on the table at the same time. At a formal dinner, however, the appetisers are served first, followed by the middle course dishes which are served according to their method of preparation. The choice and serving of the food is an art form of its own which really tests the creative ability and artistic talents of the chef. As for the choice of porcelain, the season and the dishes to be served are the decisive factors. In general, the food that is round — meatballs, sliced lotus root, for example — is served on square or rectangular plates, while "square" food is served on round plates to create a contrast. There are also serving dishes and plates that are hexagonal, semicircular, fan-shaped, leaf-shaped or in the form of shells. In most cases, the trimmings can also be eaten.

The most important flavourings in the Japanese cuisine are dried bonito and dried algae, kombu, which are a must for the clear dashi (bouillon) which is often used. A tasty miso with bean paste and wakame or nori seaweed is a quickly prepared soup. Ginger, sake and soya are common spices. Mirin (an alcoholic liquid) and rice vinegar are mixed with the rice to go with sushi, and sesame paste is often used in dips.

Each dish has its own special season, so great importance is attached to primeurs. The most important of the seasonal delicacies is osechi ryori, whose special dishes are served during the first week of the New Year. Dozens of different dishes are arranged in beautiful compositions in lacquered boxes several storeys high, which will be used numerous times during the first few days of the New Year. The traditions vary from region to region and from family to family, but fish sausage decorated with good-luck bamboo, plum and pine kernel patterns, algae rolls tied together in a bow shape with the aid of dried strips of pumpkin, chestnuts in mashed sweet potato, herring roe, grated carrots, white vinegar in a sweet dressing and pickled lotus roots should be present. Vegetables in the form of shiitake mushrooms, black radishes, lotus roots, carrots and burdock roots boiled in oil and dashi bouillon are also usually among these delicacies, as is a steaming, richly flavoured egg cream. Standard extras served with these dishes are rice cakes, either grilled or boiled in a soup, and an especially sticky kind of rice that is boiled and shaped into balls while it is still hot and then put in a wooden mortar and pestled.

Spring is welcomed with excursions to look at the cherry blossom, and restaurants serve cherry tea with beautiful flowers floating on the surface of the clear, slightly salty drink in celebration of the festival. Other spring delicacies are bamboo shoots and rapeseed flowers. Grilled eel belongs to the summer; this fish is thought to give energy to withstand the hot, humid weather. This is also the time to eat octopus, abalone, fresh fruit, vegetables and soya beans boiled in the pod and sprinkled with salt, which make a good accompaniment to a cold beer. Cold noodles with a dip of soya sauce are also popular at this time of the year. A common sight in the autumn is persimmon fruit hung up to dry. Roast chestnuts, soba noodles and mushrooms are enjoyed at this time, too, as is matsuke, a mushroom well known for its distinct aroma which is used in soups and rice dishes. During the late autumn, the year's harvest is preserved for the coming winter. The most usual method is to use miso, salt, vinegar or rice husks as preservatives. During the winter, people like to eat strips of a raw blowfish that must be prepared by a licensed chef, otherwise it may be a deadly poison. Mandarins warm up stews at this time of the year. The ritual

Japanese tea ceremony embodies the aesthetics, discipline and mystique of the country's culture. The Way of Tea (cha-no-yu) arose in the 15th century and in its original form placed great weight on the display and admiration of imported Chinese utensils. Cha-no-yu gave rise to two of the most interesting aspects of Japanese gastronomy: Japanese haute cuisine (cha kaiseki), to be served as a light meal before the tea ceremony, and wagashi, which are traditional sweets that came to play a very important role in the tea ceremony from the 16th century onwards.

The Japanese climate embraces every possibility from the chill of northern Hokkaido to the heat of the subtropical islands in southern Okinawa, which means that regional cuisines have grown up in the country. On Hokkaido, where it is not possible to cultivate rice, they have instead developed a liking for potatoes, maize, dairy products, grilled meat and salmon. Their version of Chinese noodles is often served with a pat of butter. Stews based on the fruits of the sea like crab, scallop and salmon are another of this region's specialities. The difference in what tastes people prefer can be distinguished between those living in the Kanto region (round Tokyo and Yokohama) and those living in the Kansai region (Kyoto and Osaka). In the Kansai region, miso is nearly always white while in the Kanto region it is dark or red. East and west Japan are also divided over the taste of sushi, sweets and sauces. The Kyoto region is distinguished for having preserved the haute cuisine of the old imperial court with mild, refined flavouring. Nagoya (between Tokyo and Kyoto) is known for its flat noodles and its sweet rice jelly. The fresh sardines and the mandarins from the island of Shikoku are greatly appreciated by the pilgrims who visit the Buddhist temples there. The island of Kyushu is known for its tea, fruit, fish and shellfish, but also for its Chinese and western influences. Pork dishes are popular on the islands of Okinawa, as well as desserts and sweets made from raw cane sugar, pineapple and papaya. There are many local strong drinks. Awamori is made from sweet potato. This beverage is unique for Okinawa and it is distilled, not brewed; there is a popular kind called habu after the deadly poisonous snake that lies curled up at the bottom of the bottle.

When discussing the Japanese cuisine, it is difficult to avoid mentioning the food that has become synonymous with Japanese cooking — sushi. Sushi was first eaten in Japan several hundred years ago as a way of preserving fish. Cleaned raw fish was pressed between layers of salt and rice by a heavy stone. After a couple of weeks, the stone was removed and replaced by a lighter lid; after another couple of months the fermented fish and rice were considered ready to eat. This type of sushi (narezushi) can still be found in certain restaurants in Tokyo. It is made with freshwater carp and has such a strong taste that it quite overshadows the fish's original character. In the 18th century, sushi began to be made without this fermentation process and it was served in something like its present form. It became very popular, so popular that two varieties were developed: à la Kansai from Osaka in the Kansai district and à la Edo from Tokyo (which at that time was called Edo). It was fishmongers in Osaka who developed the sushi that to begin with consisted of spiced rice mixed with other ingredients, and this was later made into small decorative, edible packets. Sushi à la Edo consisted of a select piece of fish or shellfish served on a cushion of spiced rice. It is this type that we connect in the first place with the noble art of the sushi tradition.

There are three drinks served with sushi: green tea (ocha), rice wine (sake) and beer. Green tea was originally considered to be a medicine. Since Japanese food is very salty, and the water is not clean enough to drink, large quantities of tea are consumed at mealtimes, as soon as the opportunity offers itself. Green tea is refreshing and can also be drunk along with sake. Traditionally, the Japanese only drink green tea; black tea is considered odd although both kinds of leaf come from the same bush. To make green tea, fresh leaves are steamed to destroy the enzyme that otherwise would ferment, giving the leaves the black colour that ordinary tea in China, India and England has. In the 12th century, ground tea was used by the Zen monks to keep them awake during the long nights filled with prayer. Tea-drinking had its heyday under the Buddhist religion, when ritual ceremonies for tea-drinking were developed. There is tea of many different qualities in Japan and their classification depends on the methods used for cultivation and harvesting. The most expensive and best tea comes from the fresh, young leaves of very old bushes that grow in the shade. The way tea is brewed depends on its quality; simple, coarse types need hotter water. A tea of high quality (Sencha) is often served with sushi as its taste suits raw fish very well. Japanese tea is always served steaming hot, without sugar or milk. You are supposed to slurp when you drink tea so that the air you draw in cools it. Sake is a fragrant, light yellow rice wine, slightly sweet and with a sherry-like taste. In the winter it is served hot, in the summer cold; it has an alcohol content of about 17 per cent.

Calling sake a wine is not quite right since it is produced from grain and not from fruit. The fermenting process is similar to that used for making beer. Sake goes well with all sorts of Japanese food and is suitable for all kinds of occasions. It is served in small porcelain bottles (tookuri) and poured out into small, delicate cups (sakazuki). The host pours the first glass for his guest and then the guest pours a glass for the host. It is a token of friendship to make sure that everyone has a full glass and when your glass is filled, you should raise it in thanks for this friendly gesture. It is not polite to refuse more sake, whether you feel drunk or not. If you feel obliged to say no, you should turn your glass upside down on the table, otherwise the host will continue serving more despite your protests, believing that you are merely being modest. When sake is drunk cold in the summer, it is taken in square boxes made of cedar wood (masu), an old measure for rice. A little salt is poured into one corner and the whole contents should be swallowed in one gulp. At sushi bars, regular guests have their own masu box, which they recognize by the teeth marks on it. If sake is to be good, it has to be made with new, high-quality rice and water from pure mountain streams. A special yeast is added to steamed rice, after which it is allowed to ferment. This process takes from six to eight weeks, after which it is ready to be drunk. Thus there is no maturing process or vintage sake. However, there are two different kinds of sake, one sweet and the other dry; the sweetness is natural while the dry type is produced artificially. There are also three qualities of sake: extra, which is the best sort and which is only served on special occasions; first class, which is still excellent but in reality is second class; and a popular and fairly cheap third class. The quality criteria are high for all three categories, and even the third class is very good. Sake is good in that it does not deaden the taste buds; rather, it enhances taste, but it does have an intoxicating effect.

Beer (biru) exists in many forms in Japan, all of which are excellent. Kirin has a powerful, nutlike taste, while Asahi is reminiscent of American beer and is the sweetest type. Sapporo is a lighter and more bitter beer, reminiscent of European beer. These types of beer have a refreshing effect on the body.

Table manners in Japan are very complicated, and a Westerner is not expected to have any table manners at all. Thus the Japanese will excuse almost any kind of behaviour. Here are a few details from this complicated set of table manners. Avoid eating half of a piece of food and putting the rest back on the plate (i.e. one should eat up the whole piece of food). You can pick your teeth at the table; the Japanese themselves do so. But they put their hand in front of their mouth like a fan. Do not pass food from one person to another with chopsticks. This reminds people of the Japanese funeral ceremony at which the cremated bones of the deceased are passed round the family. Use the other ends of your chopsticks, not the ends you put in your mouth, when you take food from a serving dish. You should preferably not leave any food on your plate. Japan is a country with limited resources and wasting food is considered a serious matter. The Japanese think it is offensive to have leftovers. In the old days, for example, noblemen took paper with them to the table and made small packages of the remains of a meal like fish bones and fruit stones which they put in the sleeves of their kimono and took home with them.

Mexico

The Mexican cuisine in the country today goes back several centuries. It is a mixture of many different influences that have blended together to make a style that is distinct and easy to recognise. Its basis lies in the ancient Aztec and Maya people and the raw materials they had access to. These included maize, avocado, fresh and dried beans, sweet potatoes, ordinary potatoes, tomatoes, chillies, pumpkin, turkey, duck, chocolate and an abundance of fish. The Spanish conquest introduced cattle that gave milk, cheese, meat, chickens, pigs, rice, wheat, cinnamon, cloves, black pepper, oranges, peaches and apricots. These new foodstuffs were quickly adopted by the native population, but not so that they altered their own culinary traditions; instead, they used the new items to improve and extend their own cuisine. In the 19th century, French influences came to Mexico, resulting in many pastries, cakes and puddings that were greatly appreciated.

The hallmarks of the Mexican cuisine are the aromatic, enticing smells that come from the cooking, giving an indication of strong dishes, fragrant with the spices that have been used. Not all Mexican food is strong and heavily spiced, however. Some dishes have a subtle taste, allowing the delicate taste of a good-quality product to emerge. Another hallmark of the Mexican cuisine is that the recipes are not complicated and the ingredients are not difficult to get hold of. The meals themselves should be a source of relaxation and pleasure.

Dried beans play a dominant role in this cuisine. They are served in many different ways from part of many dishes; they absorb and blend together spicy flavours and are nutritious. The most popular dried beans are the tasty, soft black beans and the sweet, soft pinto beans. The cheeses that is most widely eaten in Mexico is a granular, fairly salty white cheese called queso fresco.

Chilli plays a very important part in the Mexican cuisine and there are dozens of different kinds of chillies, ranging from the burning hot to the mild. In general, the smaller the fruit, the stronger it is. Chilli belongs to the same family as peppers and pimentos. The chilli that is used most frequently is the dried type. It requires a certain training and great competence to know how to regulate the strength of chilli in cooking. To get a dish that is not too strong, whole chillies are used, which are taken out just before the food is ready. To get medium strength, several chillies can be put in the pan, one of them halved lengthwise with the seeds taken out; this chilli is thrown away before the food is served. If strong dishes are wanted, chopped or sliced chillies are added, with or without their seeds; the more seeds that are used, the stronger the taste. Other essential spices are oregano, coriander, caraway, cloves, chilli powder, dried capsicum seed, aniseed, turmeric and parsley.

Good-quality lard is the usual fat used in this cuisine. Mexican cream is reminiscent of crème fraîche. Masa harina is a kind of flour made from maize that has been steeped and boiled in lime juice and then dried before being ground very fine. It is paler than ordinary maize flour and makes heavy dough. This flour is used to make maize tortillas and tamales. Tortilla is used as a side plate at meals or it can be a part of other dishes. The traditional sort are maize tortillas, but tortillas made of wheat flour are eaten in northern Mexico. These tortillas are more difficult to make and become brittle if they are cooked too long and are difficult to bend; they are thicker than maize tortillas. There are no fixed rules about what tortillas should be filled with, it depends on the availability of raw materials and the financial situation. Other variants of tortillas are burritos (soft, filled and rolled tortillas), enchiladas (filled, rolled tortillas covered with a sauce and heated in the oven), nachos (fried tortilla chips, topped with cheese and chilli and then grilled until the cheese melts), quesadillas (often filled with green chilli and cheese and heated), tacos (soft or crisp tortillas filled with a meat and sauce mixture, then folded or rolled, eaten in the hand), tortilla chips (tortilla cut into triangular pieces, usually served with a dip, guacamole) and tostadas (fried, flat tortillas topped with beans or meat, lettuce and cheese).

A kitchen utensil that should always be found in a Mexican kitchen is a tortilla press. This is a small, cast-iron hand press that makes it easier to roll out tortillas. Other important utensils are the Mexican mortar (moicajete) and pestle (tejolote). This mortar is a round bowl on three legs made of volcanic stone and the pestle is made of heavy, dark basalt. Traditionally, a great deal of the food is cooked in terracotta pots (cazuelas) or clay pots (ottas). These pots are rubbed with garlic to give the food a flavour and are then filled with hot water and a little oregano, thyme or bay leaf before being put in the oven on medium heat for a couple of hours. The pot is then taken out of the oven and the water is poured out before the food is cooked, when the flavours from the inside of the pot are transferred to it.

No Mexican meal is complete without at least one sauce on the table, preferably several. One tablespoonful of these sauces is taken with meat, tacos, eggs or beans, or is served as a dip for tortilla chips, or is poured over enchiladas before they are fried. Some sauces are made with raw vegetables and served at once, other sauces are boiled.

A Mexican cocktail party (coctel) is more of a culinary experience than a cocktail party is in other countries. These cocktail parties start early in the evening when the heavy midday meal has been digested, and continues until about 10 o'clock, when it is time for the evening meal. These coctels are an excuse for having a party, enjoying an endless variety of colourful dishes and feasting on an extravagant collection of appetisers and snacks.

The main meal of the day in Mexico is served early in the afternoon before the siesta. It begins with soup, which is intended to whet the appetite. There are two kinds of soup, wet soups and dry soups. Wet soups (sopas aguadas) are what we in the Western world call soups. Dry soups are (sopas secas) are rice or noodle dishes that have absorbed the bouillon or gravy they have been cooked in. But there is no rule without an exception. The famous Pozole is a dish in its own right and a national dish, a real meal, rather than an hors d'oeuvre.

What might be called the Mexican cuisine's fast food are antojitos, a word that means whims, fancies. In other words, these dishes are tasty snacks or titbits, based traditionally on maize pastry and masa. The most popular sorts are enchiladas, tacos and tamales.

The west coast of Mexico is often associated with fish and shellfish, even though these are widely available all over the country. Enseneda on the west coast even fills its tacos with lobster, shrimps and fish, and in Mazatlan, fish stuffed with shrimps and covered with fresh mayonnaise is a speciality. Ceviche, fresh, raw fish or shellfish marinated in lime juice is served everywhere in Acapulco. There are rich harvests of shrimps, oysters and mussels in the river estuaries and coastal lagoons.

Among the meat dishes, grilled meat is the most popular sort, especially beef and lamb, but this does not exclude the traditional barbacoa, which involves digging a large pit in the ground, wrapping the meat in agave leaves and baking it in the pit for eight hours before digging it out and eating it. Chicken is everyday food, while turkey and duck are saved for special occasions and family gatherings. Game birds are popular in the hunting districts like the east coast and parts of Veracruz. Traditionally, Mexicans ate wild duck, turkey, pheasant, quail, pigeon and turtle dove, while the domestic poultry came with the Spaniards. Chicken is grilled, made into stews, steamed or fried. They are served with sauces and rolled in tortillas or with enchiladas and tacos.

Vegetables are seldom served as dishes in their own right but are often disguised as a part of the main course of meat, fish or shellfish. The vegetables are cooked in cream, fried, filled or braised in a pot and covered with plenty of sauce. Salad vegetables, which there is no lack of, are often used as decoration, and fruit is often mixed with salad vegetables. Avocado and papaya, for example, make a splendid combination, especially when they are served with a lime dressing made from freshly pressed limes and olive oil.

Maize is always on the table in the form of tortillas, but it is also used in maize pudding, for example. Rice dishes are common on menus, but they are rather confusing because they are sometimes counted as soups since many rice and noodle dishes are called sopas secas. A rice dish never reaches the table without being flavoured or decorated. Rice is always cooked as pilau rice flavoured with a vegetable or meat bouillon. Beans are usually mashed or fried and the garnished.

Mexico is a country where beauty is measured by the kilo. This attitude means that there is no problem at all in feasting on cakes, desserts and last but not least drinks. Kilojoules and cholesterol are of no importance here. The drinks include the strong, burning tequila and the rich coffee liqueur khalua. The day begins with hot chocolate followed by milky coffee (café con leche) or tea. The main meal is rounded off with a cup of strong, black, spicy, sweet coffee before the siesta. There is a great variety of sweets and desserts, ranging from exotic tropical fruit mousses to ice-cream, nut cake, toffee, candy and crème caramel.

The Spaniards, led by Cortez, were the first to plant vines in Mexico. It takes fourth place among the wine-producing countries of Central and South America, but most of the wine produced is used to make brandy and vermouth. Nevertheless, Mexico has good wines made from Cabernet Sauvignon grapes. The tequila mentioned above is a carefully controlled type of spirits with a great variety of flavours which goes well with strongly spiced food. It may only be produced in the state of Jalisco. This is where the blue agave grows, which is not a cactus but a member of the daffodil family related to the amaryllis. There are more than 200 species of agave, but only the blue kind may be used for making tequila. The part which is used is the piñan (spike) which is a pineapple-like tuber that swells up when the plant is not allowed to flower. It takes about ten years before it is ready to be harvested and can then weigh up to 100 kg. The piñan is split, roasted in hot-air ovens or steamed and then pressed. The juice is allowed to ferment for a few days until a seven-percent, thick fluid is produced called pulque. This is also known as agave beer in the country and can be drunk as such. When it has reached an alcohol content of 55 per cent, it is distilled and then diluted with water to remove the agave taste. A mixto is a tequila that consists of other raw materials than agave; 49 per cent of the raw materials may consist of other sugars, usually maize and sugarcane. Tequila is preferably drunk at room temperature in a thick-bottomed glass called a caballito. Alongside this glass there is often another glass of cold sangrita, a red, slightly viscous and spicy drink made from chilli mixed with orange, tomato, lemon or lime juice. There are four classes of tequila: 1) Blanco/silver, which is bottled directly or after being kept for at most two months in stainless steel tanks; 2) Joven/gold, which is a young tequila coloured and with the edge taken off with caramel; 3) Reposado, which must be stored in oak casks for at least two months (the favourite among Mexicans); and 4) Añejo, which must be stored in oak casks for at least twelve months. There is also another tequila-like drink called mescal. This is often made from other sorts of agave than the blue sort and in other districts than Jalisco. Mezcal has a rawer taste and a more burnt character than tequila because the piñan is roasted on charcoal. It is in mescal that you may find a moth larva, never in tequila.

The Netherlands

The Netherlands is a collective name for the two countries Holland and Belgium. Holland is a major producer of vegetables and fruit that are exported but also appear on the country's table — an example is mussel soup. Another export for which Holland is famous is Edam cheese. When it is exported, it is in a round ball coated with red wax, but when it is sold at home it has another shape and is yellow. It takes its name from the town where it is made, and the same applies to Gouda cheese. Another of the country's specialities is herring, in Dutch matjes. This word means maiden and is used to describe small herrings that are not sexually mature which are put in a special decoction. This matjes herring is then sold at fish stalls and can be eaten in various ways. The really serious herring fan picks up the whole fish by the tail and lets it slip down his throat, sometimes accompanied by onion. Those who are not quite so enamoured of herring spike pieces on a stick and eat it with toast. The colonial era has left its mark on the Dutch cuisine in the form of the popular Indonesian food. Fish is also eaten in salads, and shellfish are used to make soup, mussel soup, for example. Meat is used in rather heavier dishes like meat and onion stew and pork and baked beans, which are often served with bread. A meal can start, for example, with a pineapple and wine punch, accompanied by cheese snacks, a cheese board and cheese truffles. Raspberry snow and fluffy desserts are examples of dishes served at the end of a meal. Almond biscuits may be served with the coffee.

Belgians love food; it is said that people could be fined in the old days for disorderly conduct if they were too fat. Mussels are frequently eaten, as are other fish and shellfish, like tomatoes filled with crab. These fish come from the North Sea and from lakes and watercourses. Watersooi fish soup should be seen as the country's national dish. Belgium leads the European egg consumption league, so egg dishes are very common, like egg au gratin. It is also the homeland of endives, witloof, which means white leaves. This vegetable is often eaten whole, only dipped in salt. It is also used in salads with avocado. Potatoes in all possible constellations play a dominant role in the Belgian cuisine and are the self-evident accompaniment to almost all food. The national drink is beer; apart from being drunk, it is often used in fish and meat dishes; beer-marinated Dover sole is an example. There are many different types of beer, even breakfast beers that are flavoured with cherries or raspberries.

As for wheat beer, Belgium enjoys a long tradition, as monks began to brew it in the town of Hoegaarden as early as 1455. It became very popular and at its height in the 18th century, 34 towns had so-called white beer breweries, but by the 1950s there were only four such breweries left. Pilsner and the Second World War were factors that led to this decline. For a long time it looked as though wheat beer had had its day after the last brewery in Hoegaarden closed down in 1957, but in 1966 it began to be brewed again and production has continued up to the present day. Belgian white beer also goes well with desserts that are not too sweet. In contrast to the German tradition, this white beer is made with a large percentage of unmalted wheat and is therefore sharp like wine with an elegant flavour. It is usually flavoured with coriander and Seville orange skin.

Nigeria

Nigeria (and other parts of West Africa along the coast) likes to use chilli in its cooking. Recipes from the coastal areas of the country include fish marinated in ginger, tomatoes and cayenne pepper and then boiled in peanut oil. Beans are staple foods. Okra (ladies' fingers) is very common in all kinds of dishes, from chicken soup to stews. Tropical fruits, especially bananas and coconut are important ingredients. Cooks from West Africa including Nigeria like to combine fish and meat; for example, they can fry dried fish in oil and combine it with chicken, yams, onions, chilli oil and water. The result is a fragrant and tasty stew. But there are not many dishes based on meat because it is in short supply there; however, there is a very popular meat stew.

Nigeria does not have a homogenous cuisine. Different tribes have their own favourite dishes. The Hausas in the north prefer meat kebabs, while the Ibos in the south favour rich stews made from fish shrimps, crabs, lobsters, rice and vegetables. The Yoruba tribe that lives in the middle of Nigeria likes meat stews best and there are divided opinions in the tribe about what they are best served with — mashed yam or mashed cassava. In the parts of the country where the population are mostly Muslim, they mostly eat food based on beans, sorghum and brown rice. The availability of raw materials plays a major role in what appears on the table. The period before the rains come in March and May is marked by a certain asceticism, whereas the period after the harvest in October-November enjoys an abundance of food. Since the Nigerian cuisine lacks a homogenous structure, it is difficult to choose one dish as the national dish, but okra soup is about the nearest one can come.

Other popular dishes are fufu, which has a flour-like texture with a smooth taste.

Fufu is eaten with soup. When stews are served, it is always with rice, but cooked plantains may also be served to make the meal more filling. Mia mia is a dish based on roots that look like potatoes and can be served to whet the appetite or served with stews. Jollof rice is a dish based on rice and meat or poultry, but these animal products can be replaced by vegetables as desired. The dish is given a lovely red colour with tomatoes. Aubergines and plantains are prepared and eaten together with spices and can be served with yams. Plantains are also eaten alone as appetisers; yellow bananas have a sweet taste in contrast to green plantains. Pepper soup is, as it sounds, a strong dish containing the fish that is available that day. Akara can be eaten as snacks or as appetisers; they are like brownies in appearance and colour but taste rather different, with a stronger flavour. Curry spaghetti says what it is, spaghetti spiced with curry and onions, sauté tomatoes, herbs and spices. Fried rice consists, apart from rice, of liver, carrots, peas, onion, tomatoes and a variety of spices.

Pakistan

The climate of Pakistan covers a spectrum from extreme cold to enormous heat, which provides the country with a great variety of foodstuffs. In general, Pakistan is a bread-eating country since wheat is a staple commodity. Other important grains are rice and maize. Throughout history, Pakistan has been invaded by different peoples like Babylonians, Persians, Greeks, Huns, Mongolians, Arabs, Turks and Afghans. This mix of peoples also created a complex mix of religions, languages and, of course, eating habits.

The great majority of the population live in villages and that is where you find the traditions that have been handed down from generation to generation. Hospitality is one of these traditions and all the important events in life, both sacral and profane, are occasions for festivities. In the towns that have been influenced by European customs, the eating habits have tended to shift from the traditional to the Western European. Out in the villages, however, food is still served in a simpler and traditional way; even if it is served with a spoon, the food is still eaten with or out of the right hand. A traditional Pakistani meal consists of a meat dish, a vegetable dish, bread, rice, pickles, fruit (there is an abundance of fresh fruit) and a dessert. The meat will never be pigsmeat, however, since Pakistan is a Muslim country. The most popular meat is veal, but beef and chicken are also much enjoyed. As for seafood, there is a wide variety of fish, shellfish, mussels and oysters. On account of the country's religion, no alcohol is drunk, but a drink that is enjoyed is lassi, a yoghourt whipped together with cream and with salt added — a most refreshing drink. The food in the north of the country is more meat based than in the south; various kinds of tikkas and koftas are often eaten together with bread. The dishes in the south are very hot, vegetarian curries that are often very colourful.

Poland

In matters concerning food, Poland shares with Russia the art of cooking deliciously full-bodied dishes from simple and inexpensive raw materials. Many dishes in Poland are strongly influenced by religious traditions. Sucking pig is eaten only at Easter, boiled carp only at Christmas. There are also special pastries baked for various red-letter days. A tasty soup called borscht (which originates from Poland, not Russia) is often served as a first course. In the summer, the soup is often cold, made from cucumber, beetroot, yoghourt or cream. Two examples of these soups are beetroot soup and sour cucumber soup. Pirogi and pies, both large and small, are common in Poland. The pastry can be ordinary or puff pastry and the fillings can be fish, meat, vegetables and/or edible fungi. Edible fungi are very common ingredients in many dishes; picking mushrooms is something of a popular movement. Wild berries and fruit are often used to make soups (raspberry soup, plum soup and blueberry soup) and creams, and they are also used as fillings in small pirogi. Many desserts are based on green cheese, preferably eaten with fruit and almonds. Like the Russians, the Poles are very fond of tea. Beer, mead (made from honey) and water are the usual drinks with meals. Vodka is drunk frequently, with all sorts of flavourings like caraway, seaweed and raspberry.

Portugal

Although Portugal borders on Spain in the same peninsula, the two countries have some differences in their gastronomies. Portugal has 850 km of coastline, which means that the population has developed a liking for fish and shellfish. Octopus, swordfish and tuna are frequently seen on the table. Grilled sardines are eaten almost every day. Then there is kabeljou (dried cod). It is said that a good chef should know at last 200 recipes for this fish, for example kabeljou pudding. Portugal's national dish is — kabeljou. From having been only rich man's food, kabeljou has moved down the social scale until it has now become the national dish.

Portugal's gastronomy changes with the seasons. Asparagus is a vegetable that occurs in various forms during the spring; wild asparagus is boiled and served with egg or a walnut sauce. Since Easter and Lent are in the spring, Easter cake is also eaten at this time — and chicken, of course. Chicken is party food and it should preferably have been raised on the farm. For this purpose alone, maize is grown to feed chickens. Sometimes the maize can also be ground to make flour for baking cakes. Everyone knows that it is not economically viable to breed one's own cocks and hens, but people do it just for the sake of their good taste. Before Lent it is common to have a feast with chicken on the table. All eating of meat during the six-week-long Lenten fast used to be totally forbidden, so before it started people would eat lots of meat. Nowadays not so many people care about the old rules of fasting, but the habit of feasting lives on. It is said that you should eat fibre-rich meat that sticks between your teeth for up to six weeks. The chicken is cooked with potatoes, onions, garlic, carrots, bay leaf and wine, and served with large bread rolls, salad, olives and lemons. Apart from this chicken dish, there are others like chicken in brandy, chicken fricassee pâtés chicken soup and chicken liver with apple. Hens are used for laying; cakes are not put on the table every day, but when people bake, they use scores of eggs. Eggs are also used in stews and soups, they can be poached or boiled. Egg yolks are used in a nutritious drink which is given to hard-working people; the egg yolks are whipped and mixed with a jug of wine.

Snails are a part of spring and are traditionally eaten on May 1st. These snails are garden snails, often dark brown but also white. They are boiled in water or white wine with olive oil, plenty of garlic, parsley, oregano, pepper, salt and mint. Boiled potatoes go well with them. When eating snails, you use the needle-sharp tips of the leaves of agave plants as eating tools. Portugal is known for its goat cheese, which is made principally in the spring The goats are milked twice a day and one litre of milk (roughly the amount yielded during a day) makes one small cheese, so a good deal of milk is needed. The fresh cheeses are wrapped in fresh fig leaves which give them a very pleasant flavour. Fresh goat cheese is eaten with a little honey, a few drops of port and a little newly ground pepper. One-week-old cheeses that have developed the special goat cheese taste are eaten as snacks with drinks. Older cheeses are grated over soups. Lamb is popular and widespread in the country; there is always fresh lamb or kid in the butcher's shops. This meat is prepared with, for example, white beans, in a marinade of rosemary and fried with grape must. In the summer, it is harvest time in Portugal, when figs, tomatoes, carob (St John's bread), peppers, almonds melons and so on are ripe. All the ripe fruits and vegetables are used in dishes served at this time such as St John's bread cake, stuffed figs, figs with green cheese, tomato soup, gazpacho, peppers, stuffed artichoke hearts, parsley pie and honey cake.

The autumn is Portugal is long and the winter short; some years there is no winter at all and autumn goes straight into spring. In the autumn there are dishes like mushrooms with garlic and oregano, flounder roulades with leek sauce, duck pâté in port, quince jam, pickled vegetables and spices, maize tartlets, bean stews and various marinades made from herbs. This is also the time when the olives are harvested. In the winter when the almond trees are in bloom people eat sweetmeats like orange cake with almond, grandma's macaroons and New Year truffles, and this is also the time when many animals are slaughtered, so dishes like fillet of pork with saffron, local pâtés and various shellfish are on the menu. The shellfish can be cooked with spices like ginger and they are also used with meat in various forms in stews. Spices are frequently used; coriander and lemon are used with both meat and fish. Vegetables and legumes are added to cold dishes like chickpea salad and in hot dishes like egg and tomato omelette. A common seasoning is the country's own wine. Refreshing citrus fruits like oranges are used in cakes like orange cake but also as a basis for an ice-cold sorbet and grilled in slices to go with almond cake.

You can't talk about the gastronomy of a country like Portugal without mentioning its wine production. Portugal is famous for its port, but it also produces other red and white wines. A reserva wine should have an alcohol content that is greater by at least .5 per cent than that laid down for ordinary wines. A garrafeira wine should also have the same greater alcohol content and in addition it must have been laid down for at least two years before being bottled and have had another year in the bottle. The grapes are not only used for wine production, however. They are also used in dishes like grape pie, chicken liver with grapes, roast pork with grape must and garlic soup with grapes.

Russia/The Soviet Union

The Russian cuisine (in Russian we include all the states that were previously members of the Soviet Union) has great variety, is sometimes complicated but is always interesting. In the Baltic countries in the northwest, salted fish, cabbage and all kinds of berries are popular. The Caucasian regions in the south like Georgia have developed a traditional cuisine based on lamb, shashlik and exotic fruits like pomegranates. Kasha or flour/grain foods in the form of bread pies, pastries and porridge are the basis of all meals. Russian bread has a taste all of its own which is considered delicious. There is a wide range of wheat and rye bread baked with special flour and in all shapes and sizes, from large, round loaves to small, ring-shaped rolls like bagels. Porridge is an important standing dish in the Russian diet; it can be eaten at breakfast, at lunch with fish, eggs or poultry or as a dessert with fresh fruit or fruit jelly. Traditionally, this cuisine has always featured eggs and dairy products. Since long ago, the typical Russian family has sat down at table to eat together. Bread is always served at mealtimes. The traditional dishes are those that improve if they are left to stand, so they are often prepared the day before they are to be eaten. All the food is placed on the table and the family help themselves, even on more formal occasions. Russians usually eat dinner (corresponding to our lunch) at about 1 or 2 o'clock. It will be a four-course meal consisting of zakouski (appetisers) soup and a main course followed by coffee or tea, fruit and dessert. Soup is eaten only at dinner. The evening meal is eaten at about 7 or 8 o'clock. Apart from the soup, the dishes follow the same order as at dinner. Vodka used to be served at all meals and was drunk out of thimble-sized glasses in one sweep. However, vodka consumption has decreased and, like wine, it is reserved for festive occasions. Kvass, a weak, home-brewed beer, is drunk at all meals by both adults and children. It has a low alcohol content, about 1 per cent, and is considered to be a refreshing drink which exists in several variants. You can also buy kvass in large barrels that are sold in the street. Making food, eating and drinking are popular pleasures and people spend a lot of time and energy in the kitchen preparing more festive meals. Root vegetables like potatoes, carrots, parsnips, turnips and onions are widely used in Russian cooking. When Russians have invited guests for a meal, it is important to have a jolly time; people talk, laugh and sing a lot at table. This can go on for hours, so the temperature at which food is served is not so important; many Russian dishes can be eaten either hot or cold.

Zakouski is the Russian term for appetisers or hors d'oeuvres. Fish zakouski is enormously popular and consists of various kinds of fish dishes like sliced sturgeon, smoked or salted, lightly salted salmon, fresh black or red pressed caviar, endless pickled herring dishes served with sour cream or mustard sauce, a kind of herring and meat pudding and pickled herring salads. There will also be aubergine caviar, pickled mushroom, salads of various kinds, pirogi of different sizes and with different fillings, various sausages and shellfish dishes on the table. Crayfish which is also included here is boiled without dill but with salt, garlic, bay leaf, tarragon and parsley in water.

The number one soup in Russia is borscht, which has its origin in Poland, where it is the national dish. This is a rich soup with or without meat that contains a number of vegetables of different kinds. Its beautiful red colour comes from beetroot, but there are types of borscht that do not contain beetroot but more cabbage. The further west you go in Russia, the more beetroot there is in the soup. The soup in northern Russia seldom contains beef, the Ukrainian soup usually has pork in it, while the White Russian soup may contain up to three different kinds of meat. Borscht is at its best the day after it has been made. Other Russian soups are also very rich and have many ingredients. If clear soup is served, it is never watery and almost always has something to go with it like meatballs or pirogi.

A large number of very aromatic and delicious poultry dishes come from the republics of Russia, Georgia and Ukraine. The Russian dishes are made with chicken, but game birds are also used. Two very special, tasty chicken dishes come from Georgia, one of which is called Tabaka chicken. This is a chicken that has been cut open and flattened out and is then roasted with a weight on top of it. It has a crisp skin and tastes of garlic, which is spread over it before it is roasted. The other dish is called satzivi iz kur. This chicken is boiled and a brownish sauce is made from ground walnuts which is poured over the dish when it is to be served. Salted herring is a national dish in Russia and other delicacies are smoked salmon, whitefish and sturgeon. The salmon or sturgeon are sliced up and put on a plate with a little dill as the only garnish, with a wedge of lemon to go with every slice. A little ground red pepper goes well with the smoked salmon. Although the former Soviet Union was one of the world's leading fishing nations, it was/is difficult to get hold of fresh fish as most of it is exported.

Smetana (sour cream) can be served with fish just as it is with meat dishes. The most famous meat dish with smetana is Biff Stroganoff. Sometimes Biff Stroganoff is light brown in colour with a slight taste of mustard, sometimes it is darker with a tomato taste. Another popular dish is shashlik, which can be prepared in several ways. A grilled saddle of lamb on a spit is a delicacy, but grilling small pieces if meat of various kinds is also very tasty and popular. Minced meat can also be grilled on a spit and then it is called ljulja kebab. Whether the meat to be roasted is marinated before roasting or not is only a matter of taste. A marinade can be made of wine, vinegar, oil, lemon juice or pomegranate juice. Lemon juice and pomegranate juice have both a preservative and a tenderising effect.

It is not possible to describe the Russian cuisine in all its complexity without mentioning the famous pancakes called blini. These can be made with wheat flour or a mixture of wheat flour and buckwheat flour. They are eaten with butter and sour cream in large quantities and with great delight in the so-called butter week that is celebrated in March, when they are said to symbolise the late-rising winter sun. Russian blini are always made with yeast. If you have a small frying pan about 10 cm in diameter, some fairly thick batter, you can fry them until they are about half a centimetre thick and then they are called oladi. There is another type called blinchiki — thin pancakes with a filling. These different kinds of pancakes can be served as a starter, a main course or a dessert depending on what is served with them, but it is most usual to serve them as a starter. Blini should be served hot with plenty of melted butter and the trimmings can vary: a pat of smetana, black caviar from sturgeon or red from salmon, thin slices of smoked salmon, salted herring and neon-eye fish all work perfectly. If blini are served as a dessert, blueberries that are not quite frozen are a delight.

Russia is almost as well known for its pirogi as for its blini. These pirogi are a crisp pastry with a golden-yellow surface, a delicious smell and very varied contents. They are a good way of making use of leftovers. There are many types of pirogi. A small one is called pirozhok, which is a piece of pirog eaten with soup or served as a cocktail snack. With meat soup they are stuffed with meat, and with fish soup with fish. A little pirog like this is served at teatime. The dough is rolled out in lengths that are cut into pieces, stuffed with a little filling and then folded together so that they look like small, plain, oval rolls. The largest type of pirog is baked on a baking tray or in a baking tin. It can be square, rectangular or like a loaf and it will have many layers of filling. At the bottom is a base of pastry with boiled rice or finely cut blini. Marrow from the spinal bone of sturgeon, called vesiga, is sometimes used for the first layer. On top of this come minced meat or fish and then mushrooms (preserved or fresh) and last of all slices of boiled egg. After that, it is all covered with rolled-out pastry that is pressed tightly together along the edges. The top is pricked a few times with a fork, brushed with egg and then baked in a hot oven until it is golden-brown. There are also round or half-moon shaped pirogi cooked in butter which are made with ordinary pastry stuffed with ground meat, often lamb.

Stewed fruit made into a sort of cream with potato flour is the most common and most popular dessert. It can be made of all kinds of fruit, fresh, dried or tinned, and it can also vary in consistency from really thick to as thin as lemonade. Children prefer ice-cream and Russian ice-cream is of a very high quality. Surprise ice-cream is ice-cream on a sponge base covered with jam and with a meringue top. Just before it is served, it is placed for a few minutes in a hot oven so that the meringue turns light brown. It is then flambéd at the table with brandy. As for drinks, kvass is reminiscent of small beer made from coarse rye bread that has been dried in the oven and then drowned in boiling water. After the mixture has stood and cooled for three or four hours, it is strained and syrup, yeast and wheat flour are added. It should stand over night in a warm place to ferment and then it is skimmed and filtered and poured into bottles. A few raisins should be put into each bottle before it is corked. The bottles should be kept in a cool place. Another popular drink is tea. A samovar only contains water for the tea; the tea itself is placed in a small teapot before the hot water is poured over it. When serving, you pour a little of the very strong tea from the pot into a glass, which is topped up with hot water from the samovar. Tea is often served already sweetened with a slice of lemon. With the tea you eat various types of bread with unsalted butter, biscuits and a kind of hard wheat rings, jam and candied fruit.

A Russian kitchen is well equipped with a large number of utensils and is well furnished with food. Measuring spoons and measuring cups are indispensable. When amounts of water have to be exact, weight is used instead of/ rather than volume. The saucepans are often large with tight-fitting lids and are made of stainless steel or enamel since marinades and fruit dishes should not be cooked in aluminium or copper pans. For baking there is a rolling pin and a glazing brush and a large tray for pirogi, pies and buns. There will also be a colander and a strainer for sieving various ingredients. Sharp knives for peeling vegetables and fruit and for cutting up meat are essential, as well as a good steel to sharpen them with. Wooden spoons are necessary for mixing various sorts of dough and perforated ladles for turning and lifting fried dishes from the boiling oil. When zakouski are served, the Russian housewife has a large collection of serving dishes, bowls, tureens and plates to present the food on.

After the dissolution of the Soviet Union, the region's best vineyards lay outside the borders of Russia, most of them in the Crimea and in Georgia. Before dissolution, the Soviet Union ranked as the world's fourth largest wine producer. Rkatsiteli vines are grown in almost half of the region's vineyards, the other half consisting of a mixture of various grapes. Apart from the medium-dry Russian champanski, very few Russian-produced wines are available outside the Russian border.

South Africa

It seems as though the first vines were planted in South Africa by Jan van Riebeck in 1657, when he established a colony there for the Dutch East India Company. It is uncertain which vines they were, but a not too wild guess would be that they were Muscatel and Steen grapes. The first wine was produced in 1659. In 1684, Groot Constantia grapes were planted by Simon van der Stel, and at present this is the best-known vineyard in the country. Its wines have been enjoyed very much by people like Napoleon and Bismarck. French Huguenots escaping persecution settled in South Africa in 1688 and were the leaders of a great expansion of viniculture. But in 1885, the country was struck by the phylloxera which destroyed the vines. In 1917, the vineyards were recreated, but then the prices collapsed due to overproduction. In 1978, an association was formed to look after the wine interests of the country. Today, South Africa produces about 8 million hectolitres of wine a year and wine consumption per capita is about 11 litres a year. The wines that are called Riesling are actually made from Cruchen Blanc grapes, which originated in Gascony. These grapes give a dry, aromatic wine. Pedro Luis grapes are used to make sherry. Another grape is Chenel, which is a cross between Chenin and Ugni Blanc and which gives a light, lively wine. As for red wine, Pintonage is an example of the country's originality in this field. It is a cross between Hermitage and Pinot Noir. Pintonage was created in 1925 and gained recognition when, at the Cape Wine Exhibition in 1959, it won first prize. It is not particularly full bodied but it has a very special aroma and good colour. The Pontac grape gives wines with good colour, a distinct taste and a good bouquet; it also ages well. South Africa's vineyards are located along parts of the coast, starting at Vredendal in the northwest, passing Capetown and continuing further east to Klein Karoo. The only vineyard outside this region is by the Orange River.

As the Dutch were the first to establish a base in South Africa, Capetown, which acted as a port linking Europe with the Far East in the 18th century, they have left their mark on the country's gastronomy. This can be seen in dishes like vanilla cake and spiced cakes. Other influences soon made themselves evident since the Dutch needed a workforce as their community developed. These workers came mainly from the Far East, Java, Sumatra and Malaysia. The Malaysian gastronomy that came to the country proved to be very adaptable to minor changes, so it fitted in with what the new country had to offer. This can be seen in the dish called bredie. This is for a South African what stew is for an Irishman Thinly sliced onion is fried brown in mutton fat, after which meat or fish is added and browned with the onions. Vegetables like tomatoes, beans, carrots, peas or marrow are put on top of the dish, spices like chilli, green ginger, cinnamon, garlic and cloves are added and it is all braised until the meat falls off the bone or the fish falls to pieces. Sosaties is another dish reminiscent of the Far East. It contains lamb or mutton marinated in a mixture of chopped onions, curry powder, chilli, garlic and tamarind powder. The meat is wrapped in fat and grilled, boiled or fried. The marinade is heated and reduced to a sauce which is coloured with saffron.

The way fish is preserved is also attributed to Malaysian gastronomy. When they were going on a voyage, they would dry, salt or pickle fish which they took on board the East India ships and it would last until they reached China. This technique lives on and a pickled, spiced fish served cold on a hot day is considered a delicacy. The arrival of the Huguenots has, strangely enough, had little influence on the South African cuisine, but there are still dishes in which grapes and/or wine are one of the ingredients, like chicken stuffed with grapes or meat stewed in wine. The Scots brought fruit cake, desserts, puddings and preserves. The basic recipe for these dishes was the same as in the old countries, but instead of the fruit they were used to they took the fresh fruit the new country offered. From the native population, the newcomers learnt how to use millet. It was ground and used for porridge, bread and beer. Maize was brought to South Africa from the United States and most South Africans have grown up eating corn on the cob with melted butter and maize bread.

South Africa has also adopted dishes from the Scandinavian countries in the form of hors d'oeuvres served at the beginning of a meal. Nowadays, pâté has a special status among these dishes but it has not been especially long known in the country. Before it made its entry, dishes like fish sambal or boiled bone marrow were served, together with brown bread and butter. Other hors d'oeuvres are mussels cooked with garlic, herring with apple, deep-fried anchovies and small fish pies. Soup is an old dish in South Africa which often today starts a dinner. These soups are like a thick purée in consistency and are eaten with bread. The main ingredients are almost always vegetables or root vegetables.

Spain

Various people and cultures have for a very long time settled in the Iberian peninsular, which is not unexpected considering its geography, surrounded by the Mediterranean Sea on one side and the Atlantic on the other. The Spanish cuisine reflects this great variety.

In the south of Spain, Moorish influence can be seen in foodstuffs like raisins, almonds, pine kernels, cinnamon and saffron. Compared with the French cuisine, Spanish cooking is less classical. The east coast of Spain is well known for its orange and lemon orchards. Andalusia exports olives, olive oil and sherry. Considering the climate, it is not surprising that the food is light, for example, the famous gazpacho, a cold, clear soup made with tomatoes, garlic, and oil and served with bread. This dish comes originally from Seville. Valencia is the hometown of the well-known paella, which is now served all over the country in a wide variety of forms. Along the coast, fish, shellfish and octopus are very common, especially in Galicia. Dried cod (kabeljou) is eaten mainly in Basque and sardines are typical in the whole of the north of Spain.

Until quite recently, every village was dependent on what was available nearby, so the cooking had a very local character. One example is that some villages in Galicia created a whole series of sweet and tasty chestnut dishes as they were surrounded by chestnut groves. Another example is that walnuts left their mark on the cooking wherever walnuts trees flourished, while in the south it was almonds that were used in the village cooking. Food there was quite simply seasonal. There were no oranges in the summer since they ripen in the winter, nor were there any tomatoes at Christmas since they are summer vegetables. In the autumn, there were no artichokes, since they bud in the spring. Most of the kitchens on the northern plains and in the middle of the country have an open fireplace under a large hood that leads to a chimney. A not unusual sight in these kitchens is a kettle standing on a tripod that simmers over the fire from early morning onwards. Inside the hood itself, sausages hang in rows to dry in the aromatic wood smoke. There are also well-ventilated pantries where hams and kabeljou are kept, together with vegetables like chillies, garlic, green beans (parboiled and dried), melons, lentils, onions and potatoes. Preserved fruit and home-made jam in jars stand on the shelves. On farms in the south of the country, the kitchen is separate from the living quarters so that the stove does not heat the house. Here there will be not an open fireplace but a tiled bench with a niche filled with gridirons above a charcoal fire. Coloured ceramic bowls and copper pots stand in rows on a shelf above the stove, but the place of honour among kitchen utensils is held by a brass mortar. In the outdoor larder there will be bunches of raisins, dried apricots, figs, sun-dried tomatoes and aubergines.

The daily rhythm is set by the mealtimes. Breakfast is eaten at dawn, after which people go off to work. Breakfast for workers, schoolchildren and businessmen usually consists of very milky coffee with plenty of sugar or hot chocolate and bread. People doing heavy work eat a more substantial breakfast consisting perhaps of garlic soup or porridge. Both schoolchildren and workers take a lunch box with them when they leave the house. On Sunday mornings, the whole family are gathered together and breakfast takes rather longer than on working days. A snack is eaten during the mid-morning. Supper, which is the main meal of the day, is eaten as late as 9 or 10 o'clock in the evening. When there is a red-letter day, however, this rhythm is broken and out in the country people feast all night long with chocolate and doughnuts. When it is time to walk to the saint's shrine, people take picnic baskets with them filled with special dishes. Something frequently served with a drink before a meal are tapas. These are small appetisers that originally consisted of pieces of bread with some sort of ham or cheese on them. Nowadays they can be all kinds of snacks or appetisers, such as deep-fried shrimps, stuffed mushrooms and mussels in tomato sauce.

Wine, beans, fresh vegetables, fresh fruit, fish, rice, bread and olive oil are basics in Spain's gastronomy. Traditionally, the consumption of dark meat and dairy products has been limited. Lentils and dry beans play an important role; the dry beans are usually large and white, but there are red and black beans as well. Fabes are large white beans like broad beans. A stew is not a stew without chickpeas as a basis. Black, green and brown lentils are used in soups. A finely ground, not too sweet biscuit without flavouring is used to thicken sauces. These biscuits are also eaten at breakfast. Eau de vie is used in small quantities to improve the taste of sauces; it is produced mainly in the sherry region of Jerez. Eau de vie is also used to flambé dishes. Meat and fish are sautéed in breadcrumbs using day-old bread that is crushed in a mortar. Cheese is another important food; Spain produces about 35 different kinds of cheese from cow's milk, goat's milk and sheep's milk, but some kinds are only produced on a very small scale and are never exported. The best-known sorts made in the traditional way are manchego, cabrales, burgos, idiazabal, roncal and villalón. Cheeses can be green, medium-mature or mature. Garlic is a very popular additive, especially if it has been roasted. Spain has a special kind of ham called mountain ham. This is cured but never smoked and is served raw. As for herbs, flat parsley is used a great deal, as are mint, bay leaf, celery, wild fennel, oregano, tarragon and thyme. The most common spices are aniseed (Christmas candies) cinnamon (desserts, custard), cloves (stews), coriander (marinades), cumin (meat and poultry dishes, gazpacho), nutmeg (meatballs, ground meat), pepper powder, pepper, saffron (rice, fish, potatoes) and sesame seed (pastry). Spain is famous for its sausages which are produced in a wide variety of forms.

Spain also produces wine in all the regions except Asturias. Here, cider is drunk instead. White, red and rosé wines are made and the origin of the wine is given on the labels. Rioja, Ribero del Duero, Rueda, Navarra, Rias Baixas-Galicia and Penedés are the homes of some of the best table wines in the country Wines that are to be drunk young have the year of production on the label. Wines that have been matured for at least two years in wooden casks and bottles have the word crianza on the label. A red wine labelled reserve must have been matured for at least three years, one of them in an oak cask A gran reserve is matured even longer. Apart from wine production, as already mentioned, sherry is produced mainly in Jerez (Cádiz) and montilla in Córdoba. These are produced using the solera process by which young wines are blended with older wines. They are fortified, so they have a higher alcohol content than table wines. Fino is a dry, smooth sherry that suits tapas, shellfish, ham and nuts. The sweeter oloroso, Pedron Ximénez and sweet Malaga muscatel are delicious dessert wines. Another type of wine is cava, a sparkling wine made by the champagne method. This comes mainly from Penedés-Cataluna. White wine is used mostly for cooking.

Sweden

Swedish food culture has been influenced over the years by many other countries. There has always been immigration and these immigrants brought with them their eating habits and gastronomic traditions.

The Swedish smörgåsbord, famous both in Sweden and abroad, originates from the hospitality and consideration that was shown in the old days to guests and relatives who had travelled a long way. These get-togethers were to a large extent connected with various festivals that were held during the year, either church festivals or family celebrations like weddings, christenings, funerals, as well as festivals that had to do with work on the farm, like harvest festivals. At such parties, the farm would always fill the table with everything it could produce. What had been dried, smoked, salted and stored in the larders was brought out, cut up and placed on the table. In the 17th and 18th centuries, when Sweden was a great power in Europe, it became the custom to converse with the ladies and serve Rhine wine. These new fashions spread some anxiety among the peasantry as the men were not used to conversing with the women. A glass of something was absolutely necessary to strengthen the heart before sitting down to eat. That was the reason why the men started the custom of beginning a party with schnapps. While the women took off their coats and bonnets, the men got together over a drink or two — or three or more. The snacks that went with these drinks was what the house could always provide: bread, butter, cheese and pickled herring.

Sweden has a long tradition of pickling herring in order to improve its keeping qualities, but Swedes have also eaten Baltic herring in various forms. Herring and Baltic herring are in fact the same fish, Clupea harengus. The fish caught in the Atlantic, the North Sea and along the west coast of Sweden are called herring, while those caught in the Baltic north of Kalmar on the east coast are called Baltic herring. Herrings vary in shape and size depending on where they have grown up and where they spawn. Baltic herring are normally smaller than herring. Swedes have not only pickled herring and Baltic herring, they also make gravad herring, use it in soups, smoke it (bloaters), fry it in various dishes with various seasonings and use it in au gratin dishes and puddings.

Before potatoes were introduced into Sweden in the 18th century, bread was the staple food. Different kinds of bread were baked in different parts of the country; in the south it was soft rye bread and in central Sweden it was crispbread. Barley bread was most common in Norrland. There was usually baking twice a year, at Christmas and at Midsummer, and if people baked too often, it was taken as a sign of poverty. Bread played an important part at festivals and the shape of the bread was decided by the particular festival.

Because Sweden stretches a long way from south to north, different dishes have been developed in different parts of the country and these have come to be special province dishes. Here is a sample list of such dishes and their probable province of origin, from north to south:

Lappland in the far north has soup made from snow grouse, Midnight Sun sandwiches (using tunnbröd, thin, unleavened bread) gravad salmon trout, mountain charr fried in paper, reindeer stew, dried reindeer meat, roast reindeer, cloudberry cake and blueberry porridge.

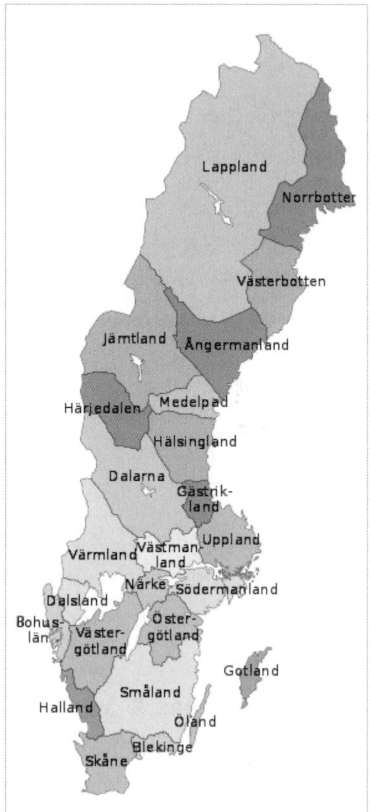

Provinces of Sweden

Norrbotten has boiled char, chanterelle soup, grilled fillets of grayling, whitefish roe (natural and fried) cold mountain fish in aspic, roast fillet of reindeer with cream sauce, blodpalt and pitepalt (blood and potato dumpling with pork), sliced, dried reindeer meat, filbunke (sour milk with crumbled crispbread), pancake tart, pancake buns, lillkaka (small cakes), old-style fresh bread, Piteå wafers, rye rusks, honey cake and snow tart.

Västerbotten has salmon soup, boiled charr, mountain whitefish gravad in dill, smoked reindeer steak with cheese and scrambled egg, reindeer offal hash, liver and delicatessen dumpling with pork, kidney dumplings with string beans in white sauce and cloudberry dessert.

Ångermanland has nätting (an eel-like fish), fermented Baltic herring au gratin, roe pudding, moose steak and various desserts made with apples and cloudberries, semolina pudding with wild berries and cloudberries (fresh or preserved).

Jämtland has game-bird soup, fried river charr with edible fungi and rice, salmon trout à la Jämtland, reindeer in various forms, boiled, salted pork with fresh cabbage, blood bread with pork sauce, Jämtland offal hash, old-style apple cake, ropy milk and oven wafers.

Härjedalen has salmon trout with cream sauce, stuffed hare, minced hare, various kinds of moose and reindeer dishes, lingon cake, various kinds of desserts made from apple and cloudberry and rosette pastry.

Medelpad has various kinds of Baltic herring dishes, offal hash, snow grouse, moose dishes, broad beans with pork, blueberry pie and apple pie.

Hälsingland has cabbage soup, Baltic herring and herring dishes (including bloaters), egg and anchovy with dill, chive and crème fraîche, Norrland offal hash, lingonberry pie, cloudberry jam and cloudberry mousse.

Gästrikland has various pike dishes, perch, sausage cake, rhubarb jelly, lingon jam, honey cake, and apple and plum cakes.

Dalarna has salted Baltic herring with lingonberry jam, charcoal burner's pancakes, pork with apple, Falu sausage, roast willow grouse, syrup potatoes, potato pancakes, rhubarb pie, egg tart, and Falu apple cake.

Västmanland has a variety of fish soups and other fish dishes, veal croquettes, moose meatballs, capercaillie, salted ham with chive sauce, roast wild duck, apple pancakes, egg and cheese cake with raisins, rusks, doughnuts and twist buns.

Uppland is yet another province that has a variety of fish dishes like perch in cream sauce, Roslagen fish stew, sea burbot in cream sauce and smelt. In addition, there are many Baltic herring dishes, pikeperch, stuffed cabbage rolls, Uppsala cream stew, blueberry and raspberry compote, boiled rye bread, fruit cream, and blueberry cake.

Södermanland also has many fish dishes. It also has Rydberg beef hash, creamed spinach, lung hash, bone-marrow pudding with wine sauce, beer posset with bread or egg and Södertälje twist buns.

Närke has many soups with rice and pork, like broad bean soup and Skogaholm cabbage soup. Other dishes are false ham hare with cream sauce, liver stew, stuffed onions, apple doughnuts, rice tart and plum cake.

Värmland has soups like Värmland bean soup and hare soup, and fish dishes like lumpfish with cream sauce, fried smelt, baked Vänern perchpike with chanterelle sauce, burbot roe, grilled bream with dill, Vänern perch with cream sauce and Värmland herring sauce. It also has roast grouse, Värmland sausage, buttermilk soup, cinnamon doughnuts, King Oscar's cake and fresh berry compote.

Västergötland has cheese soup, salmon trout simmered in cream, gravad and roast charr, crisp-fried salted herring with onion sauce, Västgöta sausage, hare, Kåvra lobster, fig pudding, pepper nuts and noble pastry rings.

Östergötland has given the Swedish cuisine green pea soup, Valdermarsvik cabbage soup, eel soup, boiled tench with dill sauce, fried herring, cream trifle, Grandma's crispy lingon cornets, and Count Spen's tart.

The island of *Gotland* has asparagus soup, lamb soup with barleycorn, Gotland fish stew, fish pudding, dried salted leg of mutton, roast shoulder of mutton with carrots in cream sauce, various rabbit dishes like marinated wild rabbit and boiled rabbit in mushroom sauce, spareribs grilled in ginger, stuffed rye buns, saffron pancakes and rhubarb pie.

The island of *Öland* has nettle cabbage, mutton soup, potato dumpling with pork, eel cakes, pork pudding, hare stew, pork doughnuts, solöga (anchovy, egg yolk, onion and capers fried in butter) and mushroom buns.

Småland has crayfish soup, huntsman's soup, cold pickled Västervik eel, hare stew, Kalmar hash au gratin, potato dumpling with pork, Småland sausage, various kinds of Småland cheesecake with cinnamon sauce, spinach pancakes, rye pastry twists, rhubarb pancakes, cream doughnuts and rice tart.

Dalsland has chicken soup, Vänern burbot soup, baked bream, fried smelt, Herrgård pikeperch, boiled, salted spareribs with parsley sauce, cabbage beef stew, vanilla pancakes, oatmeal tart and apple and rice pudding with cinnamon.

Bohuslän has a rich variety of fish and shellfish dishes such as mussel soup, Norway lobster soup, shrimps au gratin, crab in cream sauce, all sorts of plaice dishes, Dover sole, hake, boiled and gravad mackerel and stuffed fillets of whiting. Then there are dishes like apple rings fried in batter, Dalbo apple cake with raspberry jam and mashed apple roll.

Halland is another of the provinces with a maritime cuisine: salmon soup, cod stew, whiting fried in batter, grilled fillets of turbot with horse-radish mayonnaise, boiled salted herring with chive sauce, cod roe, steamed smoked eel and fish cakes fried in butter. There are also meat dishes like lamb stew, minced venison with morel sauce and kale.

Blekinge also has a maritime cuisine with clear fish soup, eel soup, fried spring herring, pork and sill, boiled fresh herring with mustard sauce, salted or grilled eel and kabeljou in various forms. Other dishes are minced meat cooked in the bag, cherry pancakes and sugar pancakes.

Finally, *Skåne* in the far south also has a maritime cuisine. A special feature of these sea dishes is eel, which is often served. Then there is black soup with giblets, sausages to go with the black soup, beer cheese, fried pork with rye bread dipped in the stock, Skåne onion dipped in stock, false boar steak, fried Skanör goose, cabbage and kale, salted goose, roast venison, bread pudding with mashed apple or chocolate, yeast pancakes and Grandma's apple cake.

The reader will doubtless have noticed that some of these province dishes appear in several of the provinces; naturally, these dishes have "migrated" in Sweden, partly because people have often moved within Sweden. Nowadays a good many province dishes are probably eaten in the "wrong" province. Apart from the dishes listed above, Swedish gastronomy has acquired new dishes thanks to increasing immigration. Anyone in Sweden who wants to eat dishes like pizza, hamburgers, kebab, dolmas or Asiatic food can do so without any difficulty at all.

Switzerland

Switzerland lies almost in the centre of Europe and because of its geographical location its cuisine has been influenced by Italy, Germany and France. The country's gastronomy is not homogenous, however, but varies from one valley to another. For example, the national dish of Switzerland, rösti, is prepared in a wide variety of ways. Switzerland is best known for its cheese. Inside the country, it is used for fondues, soufflés, omelettes, soups, pies and the like. It is also sprinkled over a wide range of dishes like cheese and bread au gratin, or bread au gratin with pear. Emmentaler is the number one cheese, with a very long history originating in the dairies in the valleys round the river Emme. Nowadays there is a competition about which cheesemaster makes the best-quality cheese, but it has not always been like that because in the old days the cheesemasters competed to make the biggest cheese. The record is supposed to be about 140 kg. Gruyère also has a very long history and is said to have existed since the 12th century. Gruyère has smaller holes in it than Emmentaler and a stronger, nut-like taste. Sbrinz is a dense, tasty cheese used more or less like Parmesan.

Cheese-making in Switzerland has remained a very small-scale business, partly because of the country's topography and mostly because of large subsidies. Since the Second World War, the Swiss Cheese Union has in principle bought up the country's whole cheese production, thus guaranteeing the dairies' sales. This system came to an end in 1999, however, and it is now an open question how the dairies will manage; so far, at any rate, the demand is greater than the supply. Switzerland produces more than 100 different kinds of cheese, but most of them can only be found locally. There is a constant demand for the cheeses that are exported thanks to their high quality. Many countries produce imitations of Swiss cheeses, but they cannot match the original products because of the unique effect that the climate and pastures have on the cows' milk. As for fish, charr, pike and perch are common and are usually fried in butter with almonds or boiled in the local wine. Coffee and cream gateaux very often round off the meals.

United Kingdom

A traditional English breakfast should preferably include porridge, juice, egg (boiled, poached, fried or scrambled) bacon, sausages, kidney, smoked fish, fried mushrooms, fried tomatoes, toast and marmalade. This will keep you going for some time, so lunch is usually somewhat lighter. Pub lunches are popular, often serving traditional dishes like fish and chips, steak and kidney pie, roast beef and Yorkshire pudding, Ploughman's lunch (bread, cheese and pickled onions) or various kinds of sandwiches, washed down by a pint of beer.)

High tea is the evening meal for some of those who eat their main meal in the middle of the day. It often includes a stew that has already been prepared or something that is quick to cook like chops or bacon in the winter or cold meat, cheese and pickles in the summer. This is always followed by a sweet, for example a pudding or a pie. The day is often rounded off with a light meal, that is, tea, coffee or hot chocolate with biscuits, cheese and a cake. Tea is drunk at all possible and impossible times. The traditional afternoon tea always includes small triangular sandwiches and rich fruit cake. The sandwiches are made a few hours in advance and filled with, for example, tomato and basil, cheese and walnuts, tuna and cucumber and egg and cress. A speciality is cream tea: scones served with strawberry jam and whipped or clotted cream.

Englishmen eat plenty of fish, plaice, cod and haddock being the most common sorts. Hake and skate are also popular and appear now and again on the table. Dover sole and salmon are other favourite fish, and the salmon is often smoked. Another favourite is kippers, herrings that have been split, cleaned, salted, dried and cold smoked before being grilled or fried with a pat of butter. Kippers can be served for breakfast or lunch. Shellfish are also very popular, and Englishmen consume quantities of cockles, mussels, crab, lobster and oysters.

The northern limit for vine cultivation runs through the British Isles. The Romans began to plant vines during their occupation and this tradition continued. During the Middle Ages, England produced a good deal of wine. The modern wine industry was born in 1952, when the first new stocks were planted. Today there are some 450 vineyards, most of them in southern England, that is, Kent, Sussex and East Anglia. The northernmost vineyard is in southern Yorkshire. The cold climate produces crisp wines that often need to be matured in the bottle to be at their best. White wines dominate the market, but a little red and rosé wine is also produced, as well as sparkling wine on a small scale.

United States of America

Wine has been produced in California since the 1850s, but this tradition came to an abrupt end with Prohibition, which lasted from 1920 to 1933. Around 1960, this region managed to return to the wine market thanks to the expansion of the vineyards and wineries in the Napa Valley. Unfortunately, this market suffered another setback with a phylloxeria attack in the 1970s and 1980s. Large sums of money have been spent on replacing these destroyed vines, and a positive effect of the wine beetle's attack was that it made room for new thinking about the type of vines that should be planted. Between 1990 and 1995, vines from this region have been adopted outside the United States, which indicates their high quality. Colombard is the most frequently used grape in California for making table wines. Another such grape is Chenin Blanc. Zinfandel is the grape that comes second in popularity, followed by Syrah and Merlot. Cabernet Sauvignon is excellent thanks to its balance between tannin and fruitiness. Chardonnay has a good balance between strength and delicacy of taste. The more complex Pinot Noir is also grown in the region.

California is not the only state that produces wine; Texas, Oregon, Washington and the eastern part of the US are also wine-producing regions. Wine production in Texas is very limited, using Chenin Blanc, closely followed by Sauvignon, Blanc, Chardonnay and Cabernet Sauvignon. Some sparkling wines are also produced. In Oregon, Pinot Noir dominates and it is for these wines that the region is best known, although Chardonnay, Riesling and Pinot Gris are also grown. Owing to the differences between vineyards, Chardonnay varies greatly from wines with a smooth, lemony taste to strong, buttery wines. In Washington, the most important district is Colombia Valley, where Chardonnay, Riesling, Sauvignon Blanc and Semillon are grown for white wine. For red wines, Cabernet Sauvignon, Merlot and a little Pinot Noir are used. The white wines from this district have a crisp, crunchy, fruity taste and the red wines have an intense taste of ripe fruit. As for the eastern states, New York State has largely replaced the native Vitis labrusca with Vitis vinifera, which has created a great interest in wines from this district. It produces both red and white wines, partly from the new vines but also from the native vines which have not been entirely abandoned.

The US also produces a native wheat beer that is drunk cold with a slice of lemon. This beer is brewed in the prairie states (Kansas, Nebraska, North Dakota, Oklahoma, South Dakota) and is a light, mild beer with a certain acidity. During the early phase of fermenting, Cluster, Eroica and Tettnang hops are added and Hallertau hops round off the mild taste. Other flavourings like raspberry emphasise the fruity taste that comes from top fermentation. Apart from wine and beer, the US produces Bourbon whiskey, the only distilled drink created in the country. This dates back to the 18th century, when the colonists imported stills from Scotland and experimented with spirit production using native products like berries, apples, carrots, potatoes, seeds, bark and molasses. Today, however, Bourbon is made from grain like maize, rye, barley or wheat. The grain used is decisive for the taste, which is mildest in maize whiskey. Yeast is added after boiling the grain and the wort is left to ferment. Some of the wort is saved to be used for the next day's production and the rest is distilled, giving a strong, quite clear whiskey. This raw whiskey is transferred to oak casks and allowed to mature for at least two years, which gives it its rich colour and mild taste. Bourbon is also used in candy, cocktails, desserts, barbecue sauces and in dishes all over the American South.

Since the US is an immigrants' continent, there are certainly as many dishes as there are peoples. Starting with breakfast, Western omelette is a favourite dish. This is an omelette filled mainly with ham and various vegetables that was invented by the pioneers travelling through the Wild West. Another favourite on the breakfast table is the pancakes that the US is famous for. American pancakes are very light and fluffy; about 2 decilitres of berries of any kind are often added to the batter before it is fried. A strawberry banana smoothie is a thick, smooth breakfast drink made of strawberries, bananas and yoghourt which is considered to go well with the rest of the breakfast.

As for starters and dips, Buffalo chicken wings are a very popular appetiser. They are very spicy, roasted chicken wings often eaten with an ice-cold beer. Other popular snacks are potato peel, stuffed mushrooms and quesadillas (a kind of tortilla from Mexico). Avocado, blue-vein cheese, spinach and onion dips are served with these appetisers. As for soups, chicken and noodle soup is one of America's absolute favourites, which is said to cure a cold as well as influenza. A popular rich and warming soup is green-pea soup with ham; this soup is often made with split peas. Fish soup from San Francisco is another favourite, a soup that is based on the tomato and fish soup made originally by Italian fishermen from the day's catch. That is why tomato sauce is still put in the soup together with mussels, shrimps, lobster, cod, wine and various spices and herbs.

Various kinds of salads are frequently served. Cobb salad is one of them, consisting of a combination of avocados, chicken, blue-vein cheese, bacon and egg. Spinach salad with a hot bacon dressing is popular on cold winter days. Other favourites are cole slaw, Caesar salad and potato and macaroni salad. There are also many different kinds of sandwiches, the most famous of which is the submarine. This is also known as hoagie, hero and grinder depending on where you are in the States. Club sandwiches are triple-decked sandwiches filled with turkey, avocado and various vegetables. An example of a toasted sandwich is tuna and melted cheese.

Turning to the main course, Creole shrimps have a special status. This is a dish from New Orleans, where peppers, onions and tomatoes form the basis of many dishes including this one. Among the fish dishes, mention should be made of baked salmon with tartar sauce, and among the many types of pizza, the one called Chicago deep dish pizza is a good example. This is a thick pan pizza covered with cheese that was created in a pizzeria in Chicago. As for meat, the famous American grilled spare ribs deserve a mention. These are first cooked in the oven and then grilled; they are often brushed with Liquid Smoke, which is a sort of essence that gives the dish a smokier taste. Beef with sauté onion, meat loaf and chilli with meat are other well-known American dishes. So are hamburgers, which can be made with different flavours or as they are; examples of flavourings are guacamole, teriyaki, bacon and cheese. Various kinds of bread like home-baked white bread, hamburger bread, blueberry muffins, maize bread and banana bread are eaten with these main dishes.

Vegetables are also eaten in a variety of forms. Indian beans, for example, are a relic from the days of the Pilgrim Fathers. It is said that it was the Indians that taught the newcomers this dish in an attempt to show them how they should make use of the native vegetables there. It is based on Lima beans. A potpourri is mixed vegetables that are first steamed and then fried. Spinach is cooked in cream while wild rice is used for pilaus with various flavours. Potatoes are not only deep-fried, they are also baked and roasted. Now for the cakes and desserts. Probably the best-known dessert is strawberry cheesecake (cheesecake with strawberry jam). Today's cheesecakes originated in a Jewish delicatessen in New York that developed a recipe for a delicious cheesecake, a rich, creamy cake on a base made from digestive biscuits. Bread pudding is a filling dessert that uses day-old bread as a base. Among the biscuits, brownies have a special place with their built-in bits of chocolate; pecan biscuits are also popular. In the pie division, mud pie, consisting of coffee ice-cream on a chocolate base, apple pie and cherry pie should be mentioned.

The West Indies

The Caribbean, or the West Indies, comprises more than 200 islands including the Greater Antilles, the Lesser Antilles, the Leeward and Windward Islands and the Virgin Islands. The Caribbean cuisine has been influenced by many parts of the world like India, Africa, Europe and China. These cuisines have become blended with Caribbean cooking over the years, resulting in a quite unique gastronomy. The food on the West Indian islands often originated from the native Arawaks who cultivated chilli, maize, garlic, yam, sweet potato, pineapple, guava and cassava. Spices were made by grinding berries, leaves and buds. A preservative was gained from cassava which was added to meat and chilli stews, and this dish is still served on the islands under the name pepper stew. But this was not the only way of preserving food that the Arawaks knew; they also preserved meat by rubbing it with spices and chilli. The meat was then cooked slowly over an open fire and the meat became dry but very tasty. This method of preserving meat is not needed today, but it has survived and is a very popular way of tenderising and flavouring meat. It has become the central feature of Caribbean food and is even used for fast food.

The slaves who were transported to the West Indies from Africa in the 17th century to be used as cheap labour took with them foodstuffs like okra, callaloo, and various kinds of beans. The West Indies were also colonised by a number of European countries at this time, among whom the French, the English and the Spanish were the most influential. These colonists brought with them foodstuffs from the whole world. The English brought rum and Worcester sauce, the French brought herbs like chive and rather more sophisticated cooking methods, while the Spanish brought onions, sugar cane, oranges, bananas and limes. The Europeans brought salted fish and salted meat, which were fed to the slaves. Because of colonial influence, many ingredients and dishes are shared by several islands, even though they often have different names and the recipes contain somewhat different ingredients. For example, chayote is also called christophene and co-cho. Chillis are used widely throughout the West Indies for their taste, in marinades, sauces, stews, salads and so on. Cakes and desserts using coconut and bananas are very popular.

In Caribbean dishes, the spicing and marinating of meat, fish and poultry is very important even though the basic ingredients may vary according to personal taste. Caribbean chefs are very particular about the appearance of dishes, so spices that may leave visible traces in the food are often wrapped in thin cloth. Whole chillis with the stalk still on them are put in stews only to flavour them and not to make the dish hot, so the stalk is removed before the dish is served. Most recipes also include fresh herbs since their taste is superior to that of dried herbs. These herbs do not necessarily have to be chopped up but the stalks are crushed before adding them so as to give more taste. Then the whole bunch of herbs is added, but this too is removed before the dish is served. The people who make food in the West Indies put their heart and soul into it and most dishes are made without any special techniques or ingredients. Most islanders use old recipes that have been handed down from generation to generation. Recipes and quantities are only used as a guide and most people add their own secret ingredients or put their own special mark on the food. The main rule is to be inventive and to adapt the recipe to suit your own taste.

A West Indian meal often begins with soups and various starters. Some of these starters are more popular than others, and dishes like banana and breadfruit crisps and maize-flour and shellfish squares are served with breakfast, lunch and dinner. A special feature of this cuisine is not to let anything go wasted. Leftovers very often form the basis of the home-made stock that gives the soups their excellent taste. The soups are usually thick, rich and nutritious, and maize-flour dumplings and vegetables like marrow and banana are often added to them. A soup that tastes best if it stands for a day before being served is the very popular pepper-stew soup. The best-known soup from the islands is probably Callalo, and every island has its own recipe for it. The whole family often helps to make the very popular maize-flour and shellfish squares. These squares are filled with a mixture of vegetables, shellfish, spices and maize-flour mash and wrapped in a banana leaf which is tied together. They are then simmered for about an hour. As for the various crisps, they are often made from plantains and breadfruit which are cut into thin slices, fried and served with a dip according to taste.

Thanks to the warm water round the West Indian islands, there are plenty of fish and shellfish to use in cooking; they include flying fish, barracuda, crayfish and crabs. Despite the abundance of fresh fish, salted cod is still a favourite because of its special taste that is important for many traditional island recipes. Fish is cooked in various ways: boiled (salted cod with fried bananas), fried (with shellfish sauce), baked in pies (with sweet potato), or as fish loaf (with coconut and pepper sauce). Fresh shellfish of various kinds are used in cooking and are cooked in many different ways, one of the most popular of which is to stuff them and bake them.

Goat, chicken, sheep and pig meat are very popular. Just like salted cod, salted beef and salted pork are used in many of the traditional dishes. Meat and poultry are often jerked, that is, marinated or very strongly spiced and dried in the sun before being cooked. The basis of the spicing is onion, herbs, chives and garlic together with any other spices available. These meat and poultry dishes are cooked, like the fish dishes, in many ways: grilled, fried (with rice and peas), stewed, grilled on spits or as fillings for vegetables like peppers. Most dishes, whether fish, shellfish, meat or poultry, are served with rice. In fact, rice is a staple food that often forms the basis of the dishes.

Vegetables are greatly loved. Usually at least two vegetables are served with every main course; they may be sweet potato, breadfruit, cho-cho, okra, papaya, plantain, callalo, yam and beans of various kinds. Salads are eaten regularly, usually with the main dish. These salads are served with a piquant or/and well-spiced dressing or sauce. If it is a pepper sauce that is served, an extra bowl of sauce is always put on the table so that anyone who wants to can help themselves. The salads often consist of potato, okra, tomato, onion, beans, callalo, nuts and tropical fruits. Sweet puddings, cakes and buns are loved by the West Indians. The desserts contain either alcohol or fruit or they are very sweet — or all three at once. The alcohol is often the local rum that can be bought at the food shops. The local fruits widely used in all sorts of desserts and especially in ice-cream are: banana, mango, papaya, guava, pineapple, soursop, and lime. Sometimes ingredients that are more usual in fancy dishes are used in these desserts, for example breadfruit and potato. Buns and rolls are often served with slices of cheese for breakfast, lunch and dinner. Common ingredients in these buns and rolls are coconut, dried fruit, spices and rum. As for drinks, fruit juice is drunk in large quantities in fruit bowls which may or may not contain alcohol.

Jewish food

Since the Jews were a people without land and since they were driven out of Palestine for many centuries, they migrated all over the world. Thus the Jewish people are spread over very many countries, making the Jewish cuisine very international and one of the most varied in the world. Despite the great geographical distances between the Jewish people, they retained their kosher laws and their festivals, on the basis of which each Jewish group developed its own cuisine under the influence of the local cuisine.

The kashrut or kosher laws distinguish the Jewish cuisine from all other cuisines. These rules are found in the Old Testament and the Talmud and form a written legal system which is often adapted to local traditions. Some people say that these early theological and symbolic rituals were created to emphasise the Jewish people's uniqueness and their covenant with God. Regardless of this, these rules determine how food is chosen and cooked in accordance with the traditions of the Jewish festivals. The combination of these factors creates a distinctive Jewish cuisine.

The food laws of the Jews indicate how important cooking and food are in their daily lives. Thus following these laws and being kosher is a lifestyle; food is chosen, prepared and combined in accordance with traditional rules and rituals.

The Torah (the Old Testament) defines animals as kosher if they are four-legged, ruminate and are cloven-hoofed. Beef, veal, sheep and goats meet these requirements. Pigs, horses, rabbits and hares, on the other hand, are not kosher. Among birds, chicken is kosher but birds of prey, newly shot pheasants and other game birds and carrion birds are not kosher.

Only domestic animals are kosher, which means that all wild animals are not kosher, whether or not they have cloven hoofs and so on. This derives from the theological explanation that man does not have the right to take an animal's life unless he is responsible for it having lived a good life, that is, it has had plenty of food, water and the like. Thus all hunting is forbidden. Certain local traditions have also influenced what is deemed to be kosher or not. Some oriental Jews do not eat geese because they are both land and sea birds, whereas goose is one of the staple foods of Eastern European Jews.

Both meat and poultry must be slaughtered in accordance with tradition by an authorised butcher (shohet). A smooth-bladed, razor-sharp knife, twice as long as the width of the animal's throat is used to cut the throat artery, the throat and the two iliac arteries. The aim is to cause the animal as little pain as possible. The claws and the skins of the hoofs of animals must also be removed, and as much blood as possible is allowed to run out in accordance with the Torah ban on consuming blood. The meat or poultry are soaked in cold water for 30 minutes in a vessel specially set aside for this purpose. The meat is then rinsed and salted with a coarse salt called kosher salt. After that, the meat is placed on a board with holes in it and allowed to run off for one hour. The salt is then shaken off and the meat is rinsed three more times and the water again is allowed to run off. The liver must be salted and grilled in the oven or over an open fire so that all blood runs off and the meat is cooked right through. The heart must be opened up so that the veins can be removed and all blood can run out before the heart is soaked and salted. It is only the forepart of four-legged animals that is kosher, so the rear cannot be eaten by Jews unless it has been made kosher by porging, that is, by taking away all the veins and sinews from this part. Since this takes a great deal of time, it is seldom done (the reason why it is so important to get all the blood out of the meat is the theological concept that the soul of a living being is in the blood and it is taboo to consume a soul). The ban on eating meat with bilk or dairy products goes back to Deuteronomy 14:21, which says that a kid shall not be seethed in its mother's milk. This ban was also a way of repudiating the pagans' way of sacrificing animals which was thought to be barbaric and unnecessarily cruel to animals. It was also considered bad for the digestion to eat meat and milk together. In some Jewish communities they wait for up to six hours after eating meat before eating any other food that contains milk. On the other hand, it is enough to wait one hour after a milk dish before eating meat again. The same applies to cheese as to meat, that is, a three-hour wait (the rules for how long you must wait before consuming meat after milk or vice versa depend on minhag, that is, local religious traditions). These rules are followed by using different utensils, pots, plates, cutlery and even dishwashers. Milk and meat must be kept separate and must not come into contact with each other in the fridge.

Generally speaking, Jews can be divided into two groups, the Ashkenazi and the Sephardi, although these groups included several other groups and in some areas they overlap. The Ashkenazi come from Central and Eastern Europe including Austria, Germany, Hungary, Poland and the former Soviet Union. However, the Soviet Union was so extensive that the cuisine of the Jewish groups that live near the Turkish and Iranian borders is very reminiscent of the cuisine of their Sephardi neighbours. The Sephardi Jews come from Spain, Portugal and the Middle East. Some Jews who were driven out of Spain and Portugal during the Spanish Inquisition went to Holland while others returned to the Middle East.

The Ashkenazi cuisine is often mild and delicate in taste. Their tasty dishes are usually based on beef and fish and sometimes lamb, with onion, pepper powder and small amounts of garlic. Parsley, dill and chive are commonly used herbs. Strong horseradish sauce is often served with meat and fish and always with gefilte fisch (a kind of Jewish fish quenelles). Pickled vegetables are very popular, often served with meals. In Germany, Hungary, Poland and the countries of the former Soviet Union, vinegar, pressed lemon or salt and sour are mixed with sugar and honey. Sweet-sour sauces are used in dishes like sweet-sour cabbage and dolmas. In Central Europe, clarified goose or chicken fat, and smoked or salted fish are staple foods. Egg noodles are the main ingredients in many traditional dishes. Barley, lentils and kasha (boiled buckwheat) are other popular staples. Fruit and vegetables are used a great deal in Ashkenazi cooking as they are neutral and can therefore be eaten with meat or milk dishes.

Characteristic features of Sephardi cooking are olive oil, lemon, garlic and fragrant, hot spices. Jews in Greece and Turkey season their dishes with fresh herbs like dill and coriander, while the Jews in North Africa prefer cumin and dried ginger. In general, most Sephardi Jews use cinnamon in both sweet and salt dishes. Sephardi Jews prefer lamb to other kinds of meat and sea fish to freshwater fish. In the Middle East and parts of the Mediterranean region, pita bread or other flat bread is eaten with olives, aubergines, zucchinis, artichokes, tomatoes, peppers and beans. Desserts do not have the same importance in the Sephardi cuisine as in the Ashkenazi. Nevertheless, sweets dishes are cooked there whose main ingredients are often egg, filo pastry with walnuts, almonds, cinnamon and rose or orange water rather than dairy products. Thanks to the French colonisation of Morocco, the French element in the cuisine there is strong and the gastronomy refined. In this Sephardi cuisine, Sephardi spices and ingredients are combined with French elegance. The dishes are spicy, aromatically flavoured with large amounts of saffron, pressed and salt-pickled lemons, garlic, coriander and the sweet spices cinnamon, nutmeg, ginger, mace and allspice. Yoghourt is used throughout the Middle East and mustard is an important ingredient in Egypt and the eastern Mediterranean.

Most Italian Jews live in the oldest parts of Rome and they often cook Sephardi dishes. These Jews also use a great deal of pasta together with rosemary, sage, basil and other traditional Italian herbs. In India, the Sephardi cuisine has been influenced by the local cuisine, where cumin, cinnamon, turmeric, cardamom, coriander seed, garlic, ginger, chilli and fresh coriander leaves are used to flavour the food. Iraqi Jews, who are the latest Jewish immigrants to India, use large amounts of fresh herbs and vegetables in their cooking, combining them with Indian spices and flavours.

The Yemeni Jews who came to Israel from Arabia in 1949 had lived isolated from the rest of the Jewish world for so long that their recipes and traditions had been preserved in their original form. These dishes are spicy, containing cumin, turmeric, garlic, fresh coriander leaves and hot chilli paste. Their recipes have been integrated into the Israeli and Jewish cuisines.

Natural rennet is normally used to make cheese and this rennet comes from the stomach of ruminating animals, usually cows. This means that most natural cheeses are not kosher, but there are cheeses made with rennet substitute which are kosher (but are milk products). The same applies to gelatine, which is made from animal bone. Kosher gelatine is made from seaweed and is a perfect substitute.

Not all fish is kosher, only scaly fish with fins, so shellfish, crustaceans, eel and fish without scales like catfish, angler fish, shark, turbot and sturgeon are not kosher. Kosher fish need not be prepared in any special way since it is neutral (compare with the preparation of meat). Since these scale and gill rules apply to everything living in the sea, all types of molluscs and shellfish are not kosher.

All food that does not contain meat, chicken or milk is considered neutral or pareve. All fruit, vegetables, egg, vegetable oil, fish and vegetable margarine can be eaten with milk or meat. Eggs, however, must be examined since traces of blood mean that the egg is fertilised and cannot therefore be used.

As for alcohol, the Jewish people have drunk wine since Biblical days and wine is still part of the Sabbath ritual and other festivals. Wine and other spirits that contain wine, fermented grape juice and/or brandy must be marked kosher. The grapes must be harvested and prepared under authorised supervision to be kosher. Wines and spirits made of grain are also allowed. The production of kosher wine was originally safeguarded by special rules. The vineyard had first to be hoed and cleared of stones before planting the vines. Mixing grain crops and vines in the same field was not allowed. The vines must be pruned with vine knives, weeds must be removed and the vineyard watered. The grapes must be harvested between thrashing time and sowing time. No second picking was allowed, as those grapes are for the poor in the community. For the same reason, a vineyard may not be harvested or pruned every seventh year, nor every fifteenth year, which is a jubilee year. After the grapes have been brought in, they are pressed, in the old days by being trodden in vats, after which the wine is clarified and bottled.

Chapter 4

The Nobel Banquets, 1901–2001

The Banquets, 1901–1910

The first Nobel Banquet was held in 1901 at the Grand Hotel, Stockholm. Those who took part in the prize presentation in the Main Hall at the Academy of Music went directly after it was over to the hotel. The table, which was already laid, was in the shape of a horseshoe, with an extension running from the middle. This arrangement could seat 130 guests. Two toasts were drunk in champagne, for the King of Sweden and for the Crown Prince, both followed by four cheers. After a speech by the Crown Prince, he raised his glass and proposed a toast for the Board of the Nobel Foundation. This was followed by a number of speeches in quick succession, ending in several toasts. When the speeches were over, the guests had finished their dessert, so they left the table and went to various smaller rooms where they formed groups that enjoyed lively conversation over the coffee. Meanwhile the dining table in the hall was cleared and where a five-course meal had been served there were now punch bowls. While the guests enjoyed this drink, conversation continued for another couple of hours until it was time to go home. Only gentlemen were present at this banquet. There is nothing but good to be said about this, the first of the Nobel Banquets, both about the meal and the room decorations.

The following year the number of guests had increased by 20, so in 1902 there were about 150 guests. The banquet was described as "a brilliant event", attended by prominent men in the field of science and from the civil service. This year, too, the table was arranged as a horseshoe with an extension in the middle. The King's health was drunk in champagne, although the King himself was not present, and this was followed by a toast to the Crown Prince (who was present). The Crown Prince's speech of thanks was succeeded by numerous toasts and speeches in swift succession. After the meal, the guests moved to smaller rooms to have coffee. After the punch bowls had been set out, the guests continued their social evening for a few hours until they left. This year, too, only men attended the banquet.

In 1903, the banquet was again held at the Grand Hotel and attended by about 230 guests, most of them representatives of Stockholm's scientific and literary worlds. The event was described as brilliant. Before the banquet, the Grand Hotel had illuminated its façade with glittering, coloured electric lights in the form of stars and garlands. The table was horseshoe shaped with four central extensions and had a background of living plants. This year there were a number of women guests, wives of the male guests, at the banquet for the first time. This time, too, a toast was proposed for the King followed by four hearty cheers. This was followed by a toast to the Crown Prince and other speeches and cheers. The Crown Prince replied to the toast and asked the ladies and gentlemen present to drink a toast in silence in memory of Alfred Nobel. Next came a series of speeches that were duly applauded. Coffee was served in smaller rooms after the meal. More speeches were given and toasts were proposed for the ladies and others with the drinks that were served in the main hall. The meal was described as simple but exquisite.

The 1904 Nobel Banquet was attended by almost 200 guests, including a large number of ladies, in the banqueting rooms at the Grand Hotel. This year the banquet began just before 10 p.m. when the Crown Prince and Princess Ingeborg arrived. The great pillared hall, where the table was laid, was tastefully decorated with green plants, garlands and flags. A horseshoe-shaped table stretched round three sides of the room and in the centre of the room there were three smaller tables. As in previous years, champagne toasts were drunk for the King and for the Crown Prince. Responding to his toast, the Crown Prince requested the guests to raise their glasses in silence in memory of Alfred Nobel. The Literature laureate then proposed a toast for the Swedish Academy. When the speeches were over, the guests had finished their meal and it was almost midnight. Coffee was taken in smaller rooms while the punch bowls were set out in adjacent rooms and the evening continued with conversation until the small hours. A toast was drunk for all those who had been involved in a Nobel Prize during the past year. That year the King and the Queen gave a banquet at the Royal Palace on December 12th to celebrate the presentation of the Nobel Prizes. Some 50 guests were invited, including of course the laureates themselves.

The 1905 Nobel Banquet took place at 5 o'clock. Almost 200 guests awaited the arrival of the royal family at the banqueting rooms at the Grand Hotel. When they had arrived, the procession into the hall began at half past five. The hall was arranged with a great horseshoe-shaped table whose sides enclosed four lines of smaller tables where only men were placed. Covers were laid for a total of 190 persons. The Chancellor of the Swedish Universities, Erik Boström (who was Chairman of the Nobel Foundation, 1900-1907) called for four cheers for the King of Sweden. Crown Prince Gustaf Adolf then requested a silent minute in memory of Alfred Nobel. This was followed by a large number of speeches, toasts and applause. For the toasts, most of the guests were upstanding (except for some Germans who remained seated). After all the speeches and toasts, the meal was over. The guests rose from the table and, taking their dinner partner's arm, went into the smaller rooms where coffee and liqueurs awaited them. Smoking was also permitted. The social evening continued until the royal family left at about 9 o'clock. About an hour later, the 1905 Nobel Banquet was over.

The 1906 Nobel Banquet was also for about 200 guests, both men and women. When the guests arrived, stewards stood at the smaller rooms where they handed over the guests' place cards. The guests then marched in procession into the banqueting hall led by the royal family. The hall glittered in white and gold in the bright electric illumination. Along the length of the hall ran a large table flanked by other tables across the room. The middle of the room was occupied by four smaller tables. All the tables were richly decorated with roses and ribbons in the Swedish colours, yellow and blue, which were arranged in winding bouquets. A toast was drunk to the King. The Duke of Skåne then proposed a toast in memory of Alfred Nobel and reminded the guests that it was ten years since the great donor had passed away. Altogether twelve speeches were held during the meal, with toasts and appreciative applause. The meal was not over until almost midnight, when the guests had been two and half hours at table. Coffee was then served, followed by punch bowls in the main hall. Both at coffee and in the main hall the mood of the guests was not weighed down by the seriousness of the earlier speeches.

No Nobel Banquet was held in 1907 owing to the death of King Oscar II. The Nobel Day that year was described as being hung with crepe and deep in royal mourning. However, the Nobel prizes were distributed at a simple and quiet ceremony in the Lecture Hall at the Royal Academy of Sciences which lasted only half an hour. Even though no official banquet was held by the Nobel Foundation, Åke Sjögren arranged a private party in honour of Rudyard Kipling, the Literature laureate that year. This was held at the Grand Hotel in Stockholm and was attended by Kipling, his wife and ten other guests.

The 1908 Nobel Banquet was held as usual at the Grand Hotel. At 6.30, the guests went into the banqueting hall, where the tables were arranged in the traditional, large horseshoe shape with three smaller tables. Covers were laid for 170 guests. This year, too, there was a mixture of men and women. The first toast, the King's toast, was proposed by Count Wachtmeister (Chairman of the Nobel Foundation, 1907–1918) who hoped that, like his father, the new king of Sweden, Gustaf V, would be a patron of the sciences in Sweden. These words were greeted with four cheers. The Crown Prince then requested silence and proposed a toast in memory of Alfred Nobel. The King and Queen of Sweden were not present at the banquet. More speeches were made during the banquet, and afterwards coffee was served in the smaller rooms where more relaxed conversation continued for a few more hours until the party broke up around midnight. The dinner was described as elegant. No music was performed. After having sat through the two-hour long prize-giving ceremony, it was a real pleasure to sit down and eat.

A Nobel Banquet was also given at the Royal Palace by the royal family. There were no speeches, but music was played by the string section of the Svea Lifeguards Band in the room called Hvita Havet (The White Ocean). Covers were laid for 65 persons. However, this was not the only dinner given besides the ordinary one at the Grand Hotel. Professor and Mrs G. Mittag-Lefler invited guests to their home in Djurgården in honour of the Physics and Chemistry laureates and their wives. Altogether some 150 people were invited. The President of the Karolinska Institute also gave a dinner in his home for the prize-winner in Medicine.

The 1909 Nobel prize-winner for Literature was the Swedish novelist Selma Lagerlöf. The banquet began soon after 7 o'clock in the Grand Hotel's banqueting hall. Following tradition, no music was played. Covers were laid for 280 guests on the usual horseshoe-shaped table. The first speeches were for the King and for Alfred Nobel, followed by speeches about each of the laureates. Then Selma Lagerlöf spoke, bringing tears to the eyes of many of the guests, who applauded her demonstratively with cheers. She was followed by the other prize-winners, who also gave speeches, all of which were greeted with applause and cheers. After two hours of speech-making, toasts and applause, the dinner was over and the guests moved to the smaller salons to have coffee. Over coffee, a student choir sang and was duly applauded. The Crown Princess then requested that they should sing "Ack, du Värmeland, du sköna" in honour of Selma Lagerlöf, which was followed by four cheers for her and further applause. The atmosphere grew even warmer when Professor Arrhenius read out a greetings telegram from one of the previous year's laureates. The telegram described how the English laureates were gathered together in London for a Nobel Banquet, which they planned to make an annual event. When this was announced, there were cheers for all the previous Nobel prize-winners.

The 1909 Banquet was praised as a whole for being a rich and happy occasion from beginning to end. It was celebrated as it always should be. There were many ladies present. This banquet was the largest so far in terms of the number of guests, not least thanks to the fact that the Literature prize-winner was Selma Lagerlöf. The speeches were short and witty in contrast to previous years, since none of the speakers had forgotten that they were at a dinner and not speaking from a rostrum.

The 1910 Nobel Banquet was the tenth, so it was described as a jubilee in the newspapers. The banquet began at 7 o'clock in the Grand Hotel. Three of the laureates were present and from the royal family, the Crown Prince and his wife, Prince Eugen and Prince Wilhelm attended. The entrance to the Grand Hotel was lit up by a bank of lights with a star at the centre. The dinner was for about 200 guests and in honour of the jubilee year more professors than usual took part in the banquet. The tables were decorated with arches of flowers. County Governor Hammarskiöld proposed the first toast in honour of the king, after which the Crown Prince proposed the toast of Alfred Nobel, which was drunk in silence. Numerous speeches and toasts followed. With the last speech, the dinner was over. The rest of the evening was a lively social gathering which was entertained by a student choir. Coffee was served in the smaller salons. There were remarkably few ladies present compared with the year before, and only one laureate's wife was present. After coffee, no further speeches were made over the refreshments but the conversation was all the more lively and the student choir entertained the guests. This year, too, the King and Queen held a banquet at the Royal Palace in connection with the prize-awarding ceremony to which some 70 guests were invited.

The Banquets, 1911–1920

The eleventh banquet in 1911 began at about 7.30 in the Grand Hotel. As in previous years, the tables were arranged in a horseshoe shape and were decorated with chrysanthemums. Some 300 guests attended, which was more than ever before. As usual, the banquet began with a toast to the King, after which the Crown Prince proposed a toast in silence in memory of Alfred Nobel. Then followed the traditional speeches and toasts and the dinner was over when the last speech had been given. The guests then moved to the smaller rooms for coffee, where they were entertained according to tradition with student songs. The banquet was over around midnight, but it took a long time for the hall to empty after this stylish and enjoyable celebration.

In 1912, for the first time, no Peace Prize was awarded. The 1912 Banquet at the Grand Hotel began at the appointed hour since the prize-giving ceremony at the Academy of Music had not taken too long and the guests could arrive in good time. It was a lively group that met up, some 175 in number, which was fewer than the previous year. As in previous years, the tables were arranged as a horseshoe with four smaller tables in the middle. They were decorated with poinsettias and Christmas roses, which were described as being unusually stylish. Since there were fewer guests than usual, there was plenty of room at the tables. In previous years, the guests had hardly had elbow room. The first toast was for the King and was greeted with cheers. Prince Wilhelm then proposed "a respectful and grateful toast" in memory of Alfred Nobel. A series of speeches and toasts were greeted with long and loud applause. The dinner was succeeded by a jolly social gathering entertained by student songs. These were performed by a male choir with a tenor soloist and reached a climax when the singers lifted up the laureates and carried them round the hall to the tones of "Sjung om studenternas lyckliga år". After coffee, the guests returned to the banqueting hall where a jovial evening began with a promenade round the refreshments table. The royal guests left at about 11 o'clock and shortly afterwards the party was over. This banquet was described as brilliant but somewhat stereotype for the Swedish guests. The foreign guests, however, had been given a happy memory of Sweden to take home with them.

In accordance with tradition, the Nobel festivities in 1913 were rounded off with a banquet in the main hall at the Grand Hotel. This year the guests had to hurry from the Academy of Music to the hotel to be in time for the banquet at 7 o'clock. Already before the banquet began there were groups of people assembled in the small salons, engaged in lively conversation. When they entered the hall at the appointed time, there was an atmosphere that promised well for this year's banquet. The tables, a large horseshoe-shaped one and four smaller tables, were decorated with an abundance of flowers dominated by chrysanthemums which gave colour to room, and epergnes wreathed in flowers and bunches of grapes. At 7.30, the guests walked into the banqueting hall. After the toast to the King, there was cheering before the toast to the great idealist and philanthropist Alfred Nobel was given in silence. The meal, both food and drink, was described as very good and the guests as well chosen. An unusually strong element of feminine beauty gave the event a particularly lively and colourful tone, which helped to give the impression that this celebration was more successful than in previous years. The menu and wine list were in French. There was an unusually good atmosphere at the dinner table. The only thing missing was music and the fanfares that had always given a special force to the tributes to the laureates. After dinner, the mood was more relaxed than usual among the guests, who were entertained by Director Gentzel's Boys Choir. When coffee in the small salons was finished and the tables were re-laid with refreshments, the party continued in the main hall until about 11.30.

There were no Nobel festivities in 1914 owing to the situation in the world, and the presentation of the awards was postponed. In order to celebrate Alfred Nobel and the 18 years that had passed since his death, a large wreath was laid on his grave by the Nobel Foundation. Similarly, in 1915 there were no festivities. The only ceremony was the laying of a wreath at the cemetery. There are no records of either a wreath-laying ceremony or of any other ceremony in 1916 and 1917. In 1918 and 1919, however, Nobel Day was celebrated by laying a wreath on Nobel's grave. A large and magnificent wreath with ribbons in the Swedish colours presented by the Nobel Foundation took the place of honour.

In 1920, it was decided that the Nobel Prizes should be celebrated twice now that the First World War was over and because it was felt that Sweden should show itself up in all its summer glory for the laureates. The first event was described as having a warm atmosphere, both at the prize-giving ceremony at the Academy of Music and at the banquet out at Hasselbacken. After the prize-giving, the guests hurried out of the Academy's rather stuffy hall to the more attractive restaurant at Hasselbacken. The banquet there was described as stately and celebrated with cordiality and summer splendour. This day was well attended by the laureates, including those who had been awarded prizes during the war years who had been invited because there had been no festivities or ceremonies during those years.

The first Nobel Day after the war followed the traditional forms at the ceremony at the Academy of Music. Dinner was served at Hasselbacken at 7.30. The festivities this year were special in that it had been a long time since the last celebration, but the number of guests, about 100, was not as many as in the years before 1914. Whether this was because there was no Literature prize awarded or because the royal family did not take part in the banquet is difficult to say. The table was laid in Hasselbacken's large assembly room and was decorated with elegant flower arrangements.

This year another ceremony was held together with a banquet on December 10, when the 1920 laureates received their prizes. The 1919 laureates were also present. The prize-awarding ceremony took place at the Academy of Music and the banquet at the Grand Hotel was again what it used to be. After the ceremony at the Academy, there was no intermission; the guests made their way directly to the Grand Hotel. The atmosphere at the banquet was good right from the start; there was a feeling that there would not be too many or too long speeches. The royal toast was proposed, followed by Prince Carl proposing a toast in memory of Alfred Nobel. Then came the usual speeches and toasts while dinner was being eaten. The dinner ended with the last speech. This was followed by a student choir who sang ballads by Carl Michael Bellman and songs from the nations represented at the banquet. The tables this year were decorated with lilies of the valley and red tulips. This was a brilliant festival in the presence of all the leading lights of Stockholm in the fields of science and literature. The King and Queen held a banquet for the laureates on December 12.

The Banquets, 1921–1930

The year 1921 was the 25th anniversary of the death of Alfred Nobel, so the Swedish flag was flown all round Stockholm. Among the people gathered at the North Cemetery were Nobel's surviving relatives, and the Nobel Foundation laid a large laurel wreath with ribbons in the Swedish colours on his grave. Two smaller wreaths, one of pink chrysanthemums and white ribbons from the Nobel family and one of ivy from Professor Nernst were also laid.

The 1921 banquet was described as brilliant. There were about 150 guests who gathered just before 7 o'clock in the small salons at the Grand Hotel to await the arrival of the guests of honour. The royal toast was proposed by Professor Schück (Chairman of the Nobel Foundation, 1918–1929), who reminded the guests that the King, when he was Crown Prince, had presented the Nobel prizes as early as 1901. After the royal toast and "Long live the King" the King's Song was sung by everyone in the hall. The Crown Prince then raised his glass and proposed a toast in memory of Alfred Nobel, which was drunk in silence. The mood soon became relaxed and the foreign guests had no reason to complain about Swedish melancholy or coolness during the dinner. In accordance with tradition, there followed a number of speeches and toasts after the silent toast in memory of Alfred Nobel. Dinner was over after the last speech and the guests returned to the salons for coffee and conversation. About 11.30 the laureates left while the other guests remained for a while.

After the ceremony at the Academy of Music in 1922, the guests gathered for the traditional banquet for the laureates at the Grand Hotel. Professor Schück proposed the royal toast and Prince Carl spoke in memory of Alfred Nobel. This was followed by the traditional speeches and toasts by the laureates, after which the dinner was over and the guests broke up to engage in the usual social intercourse.

The 1923 Nobel Banquet at the Grand Hotel was described as "traditional". Professor Schück proposed the first toast for the King and Prince Wilhelm devoted his speech to Alfred Nobel. Then followed the usual speeches and toasts and the etiquette was followed that the dinner was over when the last speech had been made. This was followed by a pleasant social evening during which a few more speeches were made. This year's banquet brought no sensations but the evening was characterised by unbroken harmony between hosts and guests.

There were no Nobel festivities in 1924 since none of the laureates were able to come to Stockholm.

The 1925 banquet was again held at the Grand Hotel after the year's break. As usual, Professor Schück proposed the King's toast and the Crown Prince proposed a silent toast in memory of Alfred Nobel. Then followed the traditional speeches and toasts and after the last speech the dinner was over and the guests enjoyed what was described as a pleasant evening together. This year the Royal Family held a banquet on December 12 for the laureates.

The 1926 Nobel Day and prize-giving ceremony was the 25th as well as being the 30th anniversary of the death of Alfred Nobel. However, this year the prize-giving ceremony was held not at the Academy of Music but for the first time at the Stockholm Concert Hall. The ceremony took longer than planned even though the speakers had been asked to be as concise as possible. It was almost 7.30 before the guests arrived at the Grand Hotel for the dinner that should have started at 7 o'clock. There were about 250 guests and the banquet was described as "usual". The first two courses had been served before Professor Schück proposed the King's toast, after which Prince Wilhelm spoke in memory of Alfred Nobel and proposed a silent toast. Then followed a main course, which grew cold as long speeches were given in honour of the laureates (etiquette required that there should be no eating and drinking during speeches). Even though the speeches were long, the guests worked their way through the menu. At last the speeches and the dinner were over. The guests had coffee and liqueurs in the small rooms and then reunited round the refreshments table in the main hall. The ladies disappeared to the Royal, where there was dancing, and soon after midnight the stately Nobel Banquet was over. This year, too, the royal family held a banquet for the laureates at the Royal Palace.

Since there are no pictures of the first banquet, a reconstruction of the 1901 menu was made for the 90th anniversary in 2001. This photograph shows what it was thought the hors d'oeuvre looked like and how it was served. This first course was always on the menu during the first 20 years of the banquet. © Photographer: Janne Hansson

This photograph shows how it was thought the poached fillet of brill with truffles and white-wine sauce was served at the Grand Hotel in 1901. © Photographer: Janne Hansson

After the poached fillet of brill, roast fillet of beef was served, which perhaps looked like this. © Photographer: Janne Hansson

The menu of the first banquet in 1901 shows that ice-cream parfait was served, but we do not know what it looked like. This is a reconstruction made when the 1901 menu was recreated for the 90th anniversary in 1991. It is a beautiful and imaginative creation, but personally I question whether the now famous caramelised N was used as early as at the first Nobel Banquet. In 1901, people were very doubtful whether the Nobel Prizes would continue. Judging by the newspaper articles about the banquet's first year, people in those days could not imagine how greatly the banquet and the prizes would rise in public estimation. In 1901, there were only 113 male guests (women were excluded) at the banquet at the Grand Hotel. © Photographer: Janne Hansson

The first Nobel Banquet after the First World War was held in the summer of 1920 at Hasselbacken. As there are no pictures of the menu preserved, it has been reconstructed. A consommé with the fresh vegetables of the season was served first. The porcelain is from Bungalow Porslin AB, Stockholm. © Photographer: Torbjörn Boström

After the first course, cold salmon was served with grilled tomatoes and haricots verts. The porcelain is from Bungalow Porslin AB, Stockholm. © Photographer: Torbjörn Boström

The 1927 prize-giving ceremony was again held in the Concert Hall. Some 250 guests took part in the banquet at the Grand Hotel, which began at 7.30 instead of the appointed time of 7 o'clock. It was described as brilliant, a first-class Nobel Banquet This year, too, the guests had finished the first two courses and the main course was already served when Professor Schück proposed the royal toast and reminded the guests that this was the King's jubilee year. The Crown Prince then proposed a silent toast in memory of Alfred Nobel. After that, the guests were left in peace until the dessert, when Archbishop Nathan Söderblom gave a speech. His witty words, first in French, then in English, then in Swedish and finally in Italian, attracted great attention as he turned to each laureate and addressed them in their own language. His international speech was greeted with warm applause, and his humorous comments were one of the reasons for the great success of the banquet. After the Archbishop's speech had been responded to by the laureates, the dinner came to an end. Another reason for the success of the banquet was that the traditional toasts were completed quickly and that there were few and well formulated speeches. The fact that the traditional order of speeches, which previously had come between or during the courses, had been abandoned in favour of placing them after the dessert was much appreciated by the guests. The guests who had been at the previous year's banquet praised the new system because then, to their great disappointment, they had not had time for enjoyable and lively conversation with their neighbours at table owing to the endless speeches. Shortly after the last speech, the guests moved to the salons, where the members of the Swedish royal family and the Swedish colleagues of the laureates devoted themselves in particular to their foreign guests of honour. This banquet was unusually successful in both culinary and organisational respects. It was also the largest since the First World War. This year, too, the royal family entertained guests at the Royal Palace on December 12.

The 1928 Nobel festivities were described as "the 27th Nobel festivities, which were celebrated in a strictly traditional form". Two records were broken, one for the number of guests and the other for dreariness. When the ceremony was held at the Academy of Music, the atmosphere was more agreeable than it was at the Concert Hall. The Academy was always decorated with flowers, flowers and yet more flowers, whereas the Concert Hall only had bunches of spruce round the rostrum. It looked more like a funeral service than a festival. So many people wanted to attend the banquet that it was not possible to take them all. In accordance with tradition, the first toasts were for the King and in memory of Alfred Nobel, followed by the usual speeches and toasts. The banquet was described as sound but without any originality. Not everyone agreed, however, claiming that the banquet was brilliant.

This year, for the first time, voices were raised for giving the Nobel celebrations a more festive mood; it ought to be moved to the Stockholm City Hall. The argument was that the City Hall would provide the festivity that was so well needed and so well deserved. It was also suggested that the prize-awarding ceremony should take place in the Blue Hall at the City Hall, and the banquet in the Golden Hall. The experts fell into two camps, one for, the other against the proposals. Some claimed that the whole affair had become a routine, others that there was nothing to complain about and that everything was fine as it was.

Despite the previous year's debate, the 1929 banquet was held at the Grand Hotel, in the Hall of Mirrors. It was attended by some 250 people. During the very first course, County Governor Hammarskiöld (Chairman of the Nobel Foundation, 1929–1947) proposed the royal toast and expressed the sympathy of all present for the Queen in her illness. The whole of the royal family joined in the applause and cheers and in the national anthem which followed. A waltz was played during the second course, and then the Crown Prince asked for silence while they drank a toast to the memory of Alfred Nobel. The dinner proceeded peacefully, at least for a while. The guests had time to praise the Grand Hotel's cooking, the menu was satisfactory even for the most demanding guest. Then came the long speeches and toasts while people were eating their way through the menu, which the hotel enriched with fruit and Madeira wine after the dessert. After the fruit and Madeira the guests broke up and moved to the salons under lively conversation. This year there were many young, lively girls at the banquet who would probably rather have listened to jazz music and danced than sit round the refreshments table drinking brandy and water. But tradition is tradition and the young ladies had to make do with a couple of hours' conversation instead.

The 1930 prize-giving ceremony took place in the Concert Hall, while the banquet was held for the first time since the Nobel festivities began not at the Grand Hotel but at the City Hall, in the Golden Hall. It was now accepted that the Golden Hall was a considerably more appropriate place than the Grand Hotel for this international event. Before the banquet, the guests, now about 350 in number, gathered in the Blue Hall. When the first trumpet fanfares were sounded, the guests marched from the Blue Hall into the Golden Hall, where they were met with tables decorated with flowers and candelabra. The banquet began at 7.30 and right from the start there was a splendid festive atmosphere. The first toast, as usual, was in honour of the King, but this was proposed after Professor Söderbaum had welcomes the laureates. The royal toast was followed by the King's Song and then the Crown Prince proposed the traditional silent toast in memory of Alfred Nobel. Speech followed speech during the dinner and during the dessert the four laureates gave speeches, each speech being signalled with a trumpet fanfare, after which the speaker's name was called out by a toastmaster. Music was played between the speeches, including ballads by Bellman and a number of Swedish folk songs of the more sentimental kind. After the dessert, coffee was drunk at the same table. The banquet came to an end around midnight. This Nobel banquet is described as being in every respect successful, well organised and the most festive so far. At last the Nobel festivities had found their right form for the second part, the banquet. Moving from the Grand Hotel to the City Hall meant that the Nobel Banquet was given the splendour, the dignity and the festivity that it deserved and it acted as a brilliant continuation of the ceremony in the Concert Hall. Never had a Nobel Banquet been enjoyed under such splendid conditions as now. This year, too, the King held a banquet at the Royal Palace for the laureates; all in all there were 90 guests, but no music was played.

The Banquets, 1931–1940

Despite the praise that the 1930 Nobel Banquet received, in 1931 the event moved back to the Grand Hotel, to the main banqueting hall there. This year there were about 200 guests. County Governor Hammarskiöld made the first speech with a toast to the King, after which Crown Prince Gustaf Adolf proposed a silent toast in memory of Alfred Nobel. The menu was in French, as is proper, and delicious foreign wines were served. The speeches and toasts followed the same pattern as before at the Grand Hotel. The reason why the banquet was on a considerably smaller scale that the year before was the Depression, but it passed off well enough. Selma Lagerlöf was there and enjoyed the menu. The dinner was rounded off with blue grapes and a large pear and coffee was served as before at the hotel. The banquet began at 7.30 and went on into the small hours.

The 1932 Nobel Day was described as being in the good old style. There was a record number of guests this year in the Hall of Mirrors at the Grand Hotel. The meal was excellent. Despite these opinions, the Grand Hotel could not manage to create anything like the same memorable experience as the banquet at the City Hall in 1930. The music was light and playful, a mixture of international tunes. The first toast was in honour of the King and was greeted with cheers and the song "Ur svenska hjärtans djup" (From the depths of the Swedish heart), after which a toast was drunk in memory of Alfred Nobel. There were no further speeches until the main course was served. When the speeches were over, the dessert and fruit were also finished and the orchestra played the Triumphal March from Aida as the guests, led by the royal family, rose from the table to move to the smaller salons.

In 1933 the banquet was held at the Royal; this year was also a jubilee year since it was 100 years since the birth of Alfred Nobel. The Winter Garden at the Royal was the second most beautiful room in Stockholm after the Golden Hall, so it was not a bad choice for the Nobel Banquet. The tables in the Winter Garden were decorated with silver candlesticks with red shades for dinner while the dining room at the Royal was turned into a salon for the after-dinner coffee. There were about 300 guests. County Governor Hammarskiöld proposed the royal toast and the King's Song was sung, but not many people seemed to join in. It was almost only the royal family themselves who sang. The Crown Prince then gave a speech in tribute to Alfred Nobel, ending in the traditional silent toast. However, it was not quite silent during the toast since the fountain was running and a bird in the ceiling began to twitter. And so began the world's most traditional banquet. A foreign guest made a sweeping gesture above his four wine glasses and pointed out with great feeling that this event was just as remarkable every time. Then the speeches began. After the main course, a blazing bombe glacée was carried in, which brought silence to the room, though not for long, as the speeches soon continued. The banquet was described as successful since no one in the whole scientific world needed to be short of conversation.

A very early photograph of the Nobel Banquet when it had moved to the Golden Hall in the Stockholm City Hall in 1936. © The Nobel Foundation

The debut for the salon was a definite success. The meal was delicious and the many speeches gave the guests a splendid appetite. The choice of the banqueting hall was described as providing as magnificent a setting for a festival as exclusive as the Nobel Banquet as could be expected. It really radiated the festive spirit and could with pride be admired by the foremost men and women in the various fields of science. When the darkness of December had fallen outside, the beautifully patina of the walls shone cheerfully and cosily and the atmosphere was that of a real Nordic Midwinter Feast. The foreign guests were captivated and said that it would be difficult to find a more beautiful room anywhere in the world. From start to finish the mood was excellent, merry in a natural way without any trace of stiffness. Music was played discreetly throughout the evening from the minstrel's gallery.

In 1934, it was time again for the Golden Hall to house the Nobel Banquet. This was described as "well directed, a beautiful setting, good speeches — and short speeches over coffee". The guests gathered in the Blue Hall and to a fanfare of trumpets they marched into the Golden Hall, where everyone found their place within two minutes thanks to the excellent seating lists before the royal family arrived. Dinner began with a fanfare of trumpets by musicians dressed in the uniforms of King Charles XII's army, and when the first course was served, the County Governor proposed the King's toast, which was followed by the silent toast in memory of Alfred Nobel. The spectacular feast continued and the mood rose with every glass of wine. The speeches were ritually introduced and rounded off with trumpet fanfares. When the last fanfare proclaimed that the speeches were over, so was the dinner. Conversation in the galleries took over. The banquet this year was described as a success despite the terrible acoustics in the Golden Hall.

The 1935 Banquet was also held in the Golden Hall. The guests, about 250 in number, gathered in the Blue Hall before going in to the Golden Hall to the tones of a festival march. This year the banquet did not end with the usual socialising in the galleries but with a colourful students' ball in the Blue Hall. In other respects the banquet was as in previous years, starting with the royal toast and the King's Song. The students' ball was felt to be a new and beautiful feature. It began with a parade of 20 student union standards led by the Swedish flag and followed by about 300 students singing the Student's Song. The students then proposed four cheers for the laureates and continued with a few more songs before the dancing began. The speeches this year were given between the dessert and the fruit.

In 1936, the banquet was held in the Golden Hall at the City Hall and was attended by about 300 guests. The hall was decorated with pink begonias and bronze candelabra and candles. County Governor Hammarskiöld opened the proceedings as usual with a toast in champagne to the King, and Prince Gustav Adolf proposed the traditional silent toast in memory of Alfred Nobel, after which the orchestra played the waltz "Mälar Queen". The Nobel Banquet consisted of many more speeches than courses, of which there were five. Nevertheless, after the toasts, the guests were able to enjoy the table's delicacies in peace before, over the dessert, speeches were held by the Swedish professors and the laureates. When these were over, the guests broke up and moved to the veranda in the Blue Hall to await the arrival of the students with their standards and songs. The students paid tribute to the laureates with a speech in English and ringing cheers. After the Student Song had been sung, the dance began with a Viennese waltz. The banquet was called "brilliant" and the respects and congratulations the students paid to the laureates added a delightful element of colour.

The 1937 Banquet was described as brilliant but held under a threatening cloud. It took place in the Golden Hall as "a still young tradition". This year, too, there were some 300 guests. The festivities began in accordance with tradition with the royal toast and the silent toast in memory of Alfred Nobel and the following ceremonies were as in previous years. This year, too, after breaking up from the table, the guests went up to the balcony to watch the students' tributes to the laureates. The ball was described not as a students' ball but as a Nobel Ball. It was evidently so successful that some of the members of the Swedish Academy promised that the festivities next year would be reformed and enlivened.

The 1938 Banquet was again held in the Golden Hall and if earlier Nobel Banquets had come to a close with the dessert and coffee, now it was felt that the event rather began then with the entry of the students and their standards and the following ball. The previous years' procedures were followed with speeches and toast. The ball this year was described as "traditional". Thus the promised reform of the programme did not take place.

The 1939 Nobel Banquet was cancelled and Nobel Day was celebrated unceremoniously with the simple laying of a wreath on Alfred Nobel's grave. The prize money was sent to the laureates by bank draft. The diplomas and medals were sent by the Swedish Foreign Office to the ambassadors in the respective lands, who then handed them over to the laureates, with the exception of the Literature prize, since the Finnish laureate was in Stockholm. A dinner was held for him and his family there.

The 1940 Nobel prizes were put aside on account of the war and were to be presented in 1941 instead. This was the first time since 1914 that the prizes had been put aside.

The Banquets, 1941–1950

As in 1940, the 1941 festivities were cancelled on acount of the situation in the world. The money that the Nobel Banquet cost, SEK 10,000, was donated in full to the Red Cross. The banquet in 1942 was also cancelled and again the money was donated in full to the Red Cross. In 1943, the Swedish government blocked the Nobel prizes although the Swedish Academy wanted to distribute them. Once again the banquet money went to the Red Cross. In 1944, the prize money was distributed but no banquet was held in Stockholm. Instead, the Swedish congratulations to the laureates were communicated by radio in a transmission to America. The Crown Prince gave the speech on the radio. A banquet was held at the Waldorf Astoria for the laureates in New York.

In 1945, it was time again for the presentations and festivities in honour of Alfred Nobel. The banquet took place at 7.30 in the Golden Hall with almost 600 guests. Thirteen tables were laid, which is probably the maximum number for the Hall. The speeches had wisely been planned to be held between courses. There were about 70 serving staff. After coffee and the dessert were over, the drinks flowed generously at the table of honour and the guests enjoyed the students' tributes to them. The party continued with lively conversation and was not over until the small hours. The whole banquet was rounded off with drinks and dancing in the Blue Hall.

The newspapers could tell their readers as early as December 10 what the menu would be. This was the first time this had happened, and it was not appreciated by the people who had been working on the menu.

The first banquet after the Second World War was held in 1945. It started with a creamy mushroom sauce. The porcelain is from Bungalow Porslin AB, Stockholm. © Photographer: Torbjörn Boström

The mushroom soup was followed by saddle of reindeer with haricots verts cooked in butter, cream sauce and lingonberries. © Photographer: Torbjörn Boström

The 1945 menu ended with a bombe glacée with no trimmings. Like the menu served in 1920 after the First World War, the 1945 menu set the pattern for the number of courses that would be served in the subsequent Nobel Banquets with the exception of 1991, that is, three. The porcelain is from Bungalow Porslin AB, Stockholm. © Photographer: Torbjörn Boström

The newspapers described the banquet and the menu as magnificent and pointed out that the laureates and their guests would be able to enjoy an unrationed menu. The laureates for both 1944 and 1945 received their prizes. The ice-cream dessert was modelled on the tower of the City hall and topped with three gilded caramel crowns. Dinner for the 400 students who sang the student songs and performed the following dancing was served in the Blue Hall. The music that accompanied the banquet itself was light Swedish music. Nothing negative was written about the first Nobel Banquet after the war; it was described as a brilliant event and a dazzling performance.

In 1946, tribute was paid to Alfred Nobel at his grave, where a large wreath was laid and speeches were given since it was 50 years since he has passed away. The newspapers wrote very little else about Nobel Day and it seems as though the banquet followed the previous year's ceremonies The place cards were the menus, to which two newly issued 20 öre and 30 öre stamps were attached. These were stamped as first-day letters and were issued to celebrate the 50th anniversary of Nobel's death. However, many of these combined menus and place cards which were a philatelic rarity were stolen from their rightful owners when they were left on the tables when the dance began in the Blue Hall.

The 1947 Banquet was also held in the Golden Hall but is mentioned only briefly. It was presented, however, in the newspapers with photographs mostly of the royal guests and the laureates. There were 640 guests. The table of honour was decorated with tall candelabras holding burning candles and impressive flower arrangements. The first toast to the King was proposed by Mr Ekeberg (Chairman of the Nobel Foundation, 1947–1960) and was followed by the silent toast in memory of Alfred Nobel proposed by the Crown Prince. The guests conversed while the meal continued but there were no speeches; the guests could enjoy the food in peace. Not until coffee and liqueurs were served were any speeches made. As usual, the meal was excellent. The banquet ended with songs performed by students and dancing.

The 1948 Banquet also had 640 guests in the Golden Hall, followed by dancing in the Blue Hall which went on until long after midnight. Otherwise the banquet proceeded as usual. The Chairman of the Nobel Foundation, Birger Ekeberg, proposed the King's toast and the Crown Prince the silent toast for Alfred Nobel. The candles on the tables, the

Crown Princess Louise and Marshal of the Realm Birger Ekeberg at the banquet in 1947. This picture is rather special since it shows the Crown Princess being served her food, which is somewhat of a breach of etiquette. At the table of honour, the food was always served on the plates of the guests of honour, but research has revealed that the Crown Princess had asked to have it from the serving dish so that she could take her food herself. The reason was that she did not eat much and thought that she was always served too large a helping. But as the Crown Princess had been brought up always to eat all the food that was on her plate, she did so. To avoid this problem, she chose to take her own food from the serving dish. © Pressens bild reflections from the mosaic-clad walls of the Hall and the blazing light of the bombe glacée in the half-darkness that came when the chandeliers were turned off must have created a scene that the guests would remember for the rest of their lives. The ice-cream dishes were decorated with masts at the top of which were small green and white electric lights that flickered. These represented radio antennas as a tribute to space research. This is the first time that children of the laureates are mentioned as guests at the banquet. The dinner was exemplary in its simplicity this year, and the banquet was described as the glorious climax of the Nobel festivities.

The 1949 Banquet was again held in the Golden Hall and followed the traditional pattern. The flower decorations consisted of begonias and chrysanthemums.

The 1950 Banquet was described as thriving and it was attended by the cream of the world's scientists when the Nobel Foundation celebrated its 50th anniversary. The banquet was held in the Blue Hall and there were 900 guests. Thus for the first time the bombe glacée parade took place down the Blue Hall stairway. This was described as "124 lit-up City Hall towers on blocks of ice of varied colours that slowly and majestically like the break-up of the ice on the Mississippi flowed down the Blue Hall stairs on the aching shoulders of the waiters and waitresses". Nine of the dishes in the parade were decorated with the text "Nobel 50 År" in ice, that is, with one letter or figure per dish. The parade was accompanied by the Swedish folk song "Där gingo tre jäntor på vägen" (There walked three lasses on the road). Some newspapers claimed that this was the first time at the Nobel Banquet that champagne was served. One newspaper wrote that for once the champagne flowed and that there were more speeches than ever before. There were in fact eleven speeches and two toasts. The banquet ended with students performing songs and dancing, which took place in the Golden Hall as the tables were laid in the Blue Hall.

The Banquets, 1951–1960

The 1951 Banquet was held in the Golden Hall and there were 700 guests. This year was the first time that the King went on from the grandiose Nobel Prize-awarding ceremony to take part in the glittering banquet in the

Golden Hall. The dinner was described as superb. This time the bombes glacées were in the form of oil rigs.

The 1952 Banquet had as many as 1,200 guests at the Stockholm City Hall; 500 of them were university students with their partners who, before the traditional student entertainment for the guests, had dinner in the Blue Hall while the other guests were seated in the Golden Hall. The table of honour was as usual placed along the longer wall with the high windows. There was not enough room in the Golden Hall, however, for some 50 of the Nobel guests, so places had also been set in the Tre Kronor gallery. The tables were decorated with begonias. The Chairman of the Nobel Foundation, Birger Ekeberg, proposed a toast in honour of the King and the King's Song was sung, after which there was the traditional silent toast in memory of Alfred Nobel. The speeches were made before and after the dinner itself, which meant that the guests had time to drink two glasses of red wine before the first course was served. This year, 400 kg of juicy southern fruits were spread around the tables in attractive piles which were to be eaten with Old Superior White Port. This fruit was the dessert and the piles of fruit were plundered during the speeches. This dessert was much appreciated and enjoyed. All the tables except the table of honour could help themselves to fruit. Both the year before and this year, the banquet ended with the students performing songs and with a ball. The festivities were described as delightful and joyful. The banquet cost 35 kronor per person, including two different wines. This year the student singers came into the Golden Hall and sang their songs while the Blue Hall was being cleared for dancing.

The 1953 Banquet required 964 covers for the guests, not including the student guests, so tables had been laid in both the Golden Hall and the Prince's Gallery. This year, too, the students marched into the Golden Hall to perform their songs, after which there was the usual dance. The student stewards and their partners who sat in the Prince's Gallery were served a simpler menu than the guests in the Golden Hall.

The 1954 Banquet required 750 covers for the guests, 74 of which were placed in the Tre Kronor Gallery and the rest in the Golden Hall. In addition, there were 620 seats for research students and university students, who were served at long tables in the Blue Hall. The tables were decorated with carnations. The banquet concluded this year as before with the student ball in the Blue Hall. The only speech before the dessert was the traditional

A photograph taken at the more modest Nobel festivities in 1956. On account of the serious situation in the world, a brunch was arranged for the laureates in the old Stock Exchange building instead of having a grand banquet — a sign that the Nobel Foundation was sensitive to the international situation. © Pressens bild royal toast followed by the King's Song. The King responded by proposing the traditional silent toast in memory of Alfred Nobel. More speeches were made, however, at the dessert.

The 1955 Banquet was lit up by 600 candles in the Golden Hall, where there were 750 guests. The tables were decorated with red carnations and fluffy mimosa. The traditional toasts opened the banquet and the speeches to the laureates were given between the main course and the dessert. The student songs and the student ball rounded off the evening. This Nobel Banquet was described as excellent and worthy of admiration.

On November 15, 1956 the newspapers reported that the Nobel Banquet in the City Hall was cancelled. Instead of the traditional celebrations, the Nobel Foundation would hold a dinner for the laureates and a very limited number of other guests. It was presumed that the prize-awarding ceremony in the Concert Hall would take place as usual. Because of the uncertain political situation in the world, it was deemed that the traditional banquet would not be in tune with the idea behind the Nobel prizes and the testator's intentions. On November 24, the newspapers wrote that the Nobel gala would be modest on account of the critical international situation and that the festivities would concentrate on the prize-giving ceremony in the Concert Hall. The traditional banquet at the Royal Palace was also cancelled. The prize-giving ceremony did in fact take place as usual while the dinner for the laureates was held at the Exchange Building. Only people including members of the royal family attended the dinner, during which 18th-century music was played. The meal was followed by the traditional students' tributes to the laureates. It was described as a private and intimate event for the laureates. The dishes included cloudberries from Lapland which were studied closely by ten happy and curious children of the laureates. They thought the cloudberries were strange but tasty. Despite previous reports that the banquet at the Royal Palace would be cancelled, it was in fact held as a simple dinner in the King's private dining room, which the laureates' children were not allowed to attend. Only 50 guests were invited whereas very many more would normally be invited to a Nobel banquet at the palace in Karl XI's Gallery.

The Nobel Banquet returned in 1957 to the Golden Hall with 700 guests. The guests of honour and the royal family made their entry to the accompaniment of trumpet fanfares. The toast to the King opened the banquet and the King replied by proposing the silent toast in memory of

Alfred Nobel. The meal was not interrupted by further speeches, but was accompanied by a background of music which was, however, difficult to hear. The speeches did not come until the candles had burnt down so low that the wax was beginning to drip onto the tablecloths. The banquet ended with the students' songs and the ball. The dancing began at about 11 o'clock in the Blue Hall and continued for several hours.

The 1958 Banquet was held in the Golden Hall with about 730 guests, but with the students included there were about 1,200. As before, the tables consisted of the table of honour and 19 smaller tables. In accordance with tradition, the royal toast came first, followed by the King's Song and the silent toast in memory of Alfred Nobel proposed by the King. The meal could begin. The guests were served with almost mechanical precision against a background of light music. The speeches were given after the dessert. The evening ended with the usual student songs and the ball in the Blue Hall.

All that was written in the newspapers about the 1959 Banquet was that there were about 700 guests and that it was held in the Golden Hall. The youngest guest was a nine-year-old boy, the son of one of the laureates.

The 1960 Banquet was reported only very briefly in the newspapers, possibly because the whole event was televised. The banquet was held in the Golden Hall while the actual prize-giving took place in a simple ceremony at the offices of the Nobel Foundation in Sturegatan.

The Banquets, 1961–1970

No articles at all were published in the newspapers about the 1961 Banquet.

The 1962 Banquet in the Golden Hall was attended by 822 guests. Because of the large number of guests, tables were also laid outside the Golden Hall; 124 of the 822 guests sat at the table of honour. The flowers on the tables were yellow mimosa and red carnations. As always, the banquet began with the royal toast and the King responded with the silent toast in memory of Alfred Nobel, after which there were no other speeches until the dinner was over. They began when coffee was served, which became the model for all other banquets with speeches. The meal required 40,000 pieces of porcelain and 600 candles and lasted three and a half hours. After

The guests of honour enter the banquet hall in 1963. Before these guests had arrived, the other guests had been allowed in and had found their places. In this photograph, Queen Louise and her partner make their entrance while the other guests stand up. Note that the wine had already been poured into the glasses before the banquet began. © Aftonbladet bild

Dr Agnon's cover at the banquet in the Blue Hall in 1966. His cover had plates of glass instead of porcelain because he was of Jewish birth and followed the kosher laws. © Aftonbladet bild

Parts of Dr Agnon's menu at the banquet in 1966 displayed by a happy waiter. The avocado was presented and served on glass plates in accordance with the Jewish kosher laws. Judging by newspaper reports, avocado was a novelty in Sweden at that time. © Aftonbladet bild two hours, the first candles had burnt out. The university students who ate in the Blue Hall were served a simpler menu than the guests in the Golden Hall, consisting of an open salted salmon sandwich and moose steak with blackberry jelly. The banquet ended with the students' songs (which were more magnificent than usual) and dancing in the Blue Hall. During the evening a refreshments table was also set up. This year the food was judged to be excellent.

The 1963 Banquet was also attended by about 800 guests in the Golden Hall. It began with the royal toast followed by the King's Song, after which the Regent raised his glass in a toast to the memory of Alfred Nobel. The banquet ended with the ball in the Blue Hall. The speeches this year began two hours after it had started, that is, over coffee. This was the first year that a grandchild of a laureate was a guest at the banquet.

Again the banquet in 1964 took place in the Golden Hall and the guests numbered about 800 plus 430 students who ate in the Blue Hall. There were 162 waiters and waitresses serving in the Golden Hall. The banquet was rounded off by the ball in the Blue Hall.

There was no information in the newspapers about the 1965 Nobel Banquet.

The 1966 Banquet was also held in the Golden Hall with 783 guests, while the Blue Hall had 458 student guests. While the guests in the Golden Hall drank champagne, the students drank schnapps in the Blue Hall. This year one of the laureates was of Jewish birth, so his menu differed somewhat from that of the other guests. It consisted of avocado, a vegetable dish, ice-cream and biscuits, to all of which kosher wines were served. The food was not served on porcelain but on glass plates, which are considered cleaner according to Jewish custom. The City Hall had also bought new pots and pans, ladles and so on to use for the Jewish laureate's dishes. because, according to Jewish custom, it is nor permitted to prepare meat and dairy products in the same vessels. The utensils used for preparing and serving the food were then washed at the City Hall and sent to the Royal Palace kitchen to be used at the Royal Banquet.

The banquet opened with the royal toast followed by the toast in memory of Alfred Nobel. After that, there were no speeches until the very end of the meal. A breach of etiquette was committed by the Head of the Cancer Register when he lit his first cigarette before any food or drink had been served. A journalist wrote that he missed the etiquette of earlier Nobel Banquets when no cigarettes or cigars were produced until after the ice-cream. The menu served in the Golden Hall was described as "magnificent". The students' menu was simpler, consisting of a canapé of mustard-marinated crab and glazed, cold turkey with Waldorf salad and Cumberland sauce. The drinks were schnapps, beer, wine and mineral water. The banquet concluded with the students' ball, and when it began in the Blue Hall, refreshments were served in the Golden Hall. The 200 singers in the Stockholm student union's choir never ate at the Nobel banquet, but the City Hall furnished them with drinks. A hundred and forty waiters and waitresses as well as cooks and other kitchen staff were needed to ensure that the banquet went off well without any mishaps or delays.

At the banquet the following year there were 796 guests in the Golden Hall plus 450 students who ate in the Blue Hall. The evening ended with dancing in the Blue Hall and a refreshment buffet in the Golden Hall. The present king of Sweden attended for the first time as the Crown Prince.

1968 was the first year certain university students protested in public that the Nobel Banquet should be boycotted. They felt that it was out of keeping with the times and that the festivities had been carried to excess. They also reacted against the methods the Nobel Foundations used to select the guests and against the criteria for dress; for example, the student caps that had to be worn should be pure white (in other words, they couldn't wear a wine-stained cap). The students were not unanimous about this proposed boycott and voices were raised that the students who were on the list for the highly coveted tickets should be allowed to go if they wanted to, otherwise it would be a form of oppression. Those students who wanted to attend argued that they would only have this one chance to go to a Nobel Banquet.

The tables in the Golden Hall were decorated with white and yellow chrysanthemums and accommodated 796 guests. After the guests had entered the Golden Hall, they had to await the King's arrival while seated at the tables, where the champagne glasses already stood filled.

Despite the students' protest, they paid tribute to the laureates with both speeches and songs. The banquet ended with dancing in the Blue Hall, where the students had their own table. A refreshments table there served whisky and other refreshments. The man who conducted the student choir this year had taken part in the celebrations since 1912, when he himself was a student singer. At that time, it cost students two kronor to have dinner with a schnapps and a glass of Swedish punsch at the Grand Hotel, whose manager was Wilhelmina Skog. This year the cost was 50 kronor per student.

The toastmaster had been the same person since 1918, so this was his 50th anniversary. The day before he had been decorated by the King with the Pro Patria medal for faithful service to his country. The toastmaster's duties were to speak to the guests before the arrival of the King so that they knew what to do during and after the meal, to announce the speakers, to give discreet signals to the conductor of the orchestra that it was time to play, to signal to the restaurateur that it was time to bring on the serving staff with the dishes and so on. Before this toastmaster had been appointed, there had been no such person, so the guests at the table of honour themselves had had to tap their glasses for silence and were able to go on for as long as they liked, which caused trouble at the other tables, where they had not yet been served or had not yet had time to eat.

In 1969, the Nobel Prize in Economic Sciences was awarded for the first time. The Bank of Sweden established this prize in memory of Alfred Nobel. There were 796 guests present at the banquet in the Golden Hall. Now it was not only the champagne but also the hors d'oeuvre that were already placed on the tables when the guests sat down. The first course consisted of salmon, for the first time according to the restaurateur at that time. Apart from the guests in the Golden Hall there were 429 university students in the Blue Hall. As usual, the banquet began with the royal toast followed by the King's Song. The King responded by proposing the silent toast in memory of Alfred Nobel. The banquet ended with dancing in the Blue Hall. About 300 bottles of champagne, 350 bottles of red wine and spirits and soft drinks costing 10,000 kronor were served, but by midnight the drinks had run out. Alice Babs sang a solo song when the students gave their performance and she was a great success.

Now it was 1970 and the banquet attracted more guests than ever before. The reason was that there were two Swedish laureates who would be present together with their relatives and friends. Altogether there were 836 guests in the Golden Hall and this year there was a new toastmaster, the second in the history of the banquets. The guests began to gather and pour into the Blue Hall as early as 6.30 to the strains of organ music. A quarter of an hour later the toastmaster's voice echoed through the Blue Hall, telling the guests to take their places in the Golden Hall. There they found 27 tables decorated with red carnations and mimosa. The champagne and water were already poured out. The guests were to await the arrival of the royal party standing by their chairs and not seated, as they were the year before. The distance between each cover/plate should be 40 cm, which was measured with a measuring stick. At the last moment, however, this had to be changed as room had to be made for two extra guests at the table of honour. Thus it was only 38.5 cm between the covers that year. Almost everything was ready by 6.45, 13 whole minutes delayed, and according to the protocol the guests should be seated at 7 o'clock. However, the King arrived at 7.13, 13 minutes delayed. The royal toast was proposed and at 7.16 they sang the King's Song, which the King replied to with the silent toast in memory of Alfred Nobel. Two minutes later, the sauce was served and after another three minutes, the first course was brought in. When it was time to serve the main course, the staff had succeeded in catching up seven of the thirteen minutes lost in the delay. Since the banquet was held in the Golden Hall, both the kitchen and the serving personnel were working from the small areas of 21 sq m. each behind the Golden Hall. At 8 o'clock precisely, the main course was served, at 8.30 the wine glasses were refilled and 30 seconds later the water glasses were refilled. When the main course was over and the plates were changed, the staff had managed to catch up another five minutes. After the dessert, there was coffee and liqueurs, but before coffee the laureates gave their speeches. The reason for placing the speeches of thanks after the dessert is that it is a cold dish; previously the speeches were held throughout the whole dinner, which meant that the hot food got cold before the guests could eat it. At 9.40, the guests broke up and by then most of the dishes had already been taken out to the washing-up room. The tables now had to be changed while the guests were standing out on the balcony of the Blue Hall listening to the students' songs. The tables had to be prepared for the refreshments buffet since the ball began at 22.10 and refreshments would be needed. The menu was described as exotically Swedish or Scandinavian. At 22.10 exactly, the first Viennese waltz began in the Blue Hall.

For the guests it had been an unforgettable experience; for the serving staff, reinforced by some 30 16-year-old catering students, and the kitchen staff it was a routine event, just one more banquet among hundreds, nothing more than that, in spite of the delay caused by the late arrival of the royal party.

The Banquets, 1971–1980

In 1971, the Concert Hall was closed for repairs, so the prize-awarding ceremony took place at the Philadelphia Church in Stockholm. This year there were also alarming reports that the City Hall kitchen was in such poor shape that it was threatened with being closed down for sanitary reasons. It was not absolutely clear that the Nobel banquet food could be prepared there. Nevertheless, the banquet was served in the Golden Hall when the present King stood in for his grandfather who did not attend the banquet, only the prize-awarding ceremony. This year, 800 guests attended the banquet. The restaurateur had 19 detailed orders to give his serving staff, the first and probably the most important one being: "Remember, no serving or clearing can take place until His Royal Highness the Crown Prince is served. Everything is to be done after signals by the undersigned, who will be standing behind His Royal Highness's chair, and we ask you to work quickly and efficiently as usual. It is important that we keep to the timetable — and one more thing, help each other if you see anyone getting behind." This year they caught up with not just 13 minutes but a whole half-hour's delay. 280 bottles of champagne were emptied, 20 chefs worked in the kitchen. The banquet was rounded off as usual with the ball in the Blue Hall. The kitchen had been granted permission by the Health Authorities to prepare the Nobel Banquet in the City Hall only 14 days before it took place.

The 1972 prize-awarding ceremony took place at St. Erik's Exhibition Hall in Älvsjö since the Concert Hall was not yet restored, and the church service was held at the Philadelphia Church. The present King presented the prizes for the first time. The Royal Family did not take part in the Nobel Banquet as there was three weeks' court mourning for Princess Sibylla who had died. The banquet at the Royal Palace was also cancelled. 800 guests were invited to the banquet in the Golden Hall. The laureates' children and grandchildren had their own table on the balcony in the Blue Hall while the students had their dinner below.

At 6.59, the guests in the Blue Hall were asked to take their seats in the Golden Hall, accompanied by trumpet fanfares and organ music. At 7.12, the guests of honour began their entry, introduced by a trumpet fanfare. At 7.19 there was a new fanfare when the Chairman of the Nobel Foundation proposed the royal toast. One minute later, there was another fanfare to summon Olof Palme in his capacity as Prime Minister and in the absence of the King and the Crown Prince to propose the toast for Alfred Nobel. At 7.22, the first course was carried in to the accompaniment of music, at 8.57, the Stockholm Students' Choir and the Court Singer Alice "Babs" Sjöblom began their concert, at 9.12, coffee was served with liqueurs (Renault Cognac or Benedictine), at 9.27, the Deputy Chairman of the Nobel Foundation gave a speech and ten minutes later the first laureate began his speech, followed by fanfares and speeches about every four minutes. Thereafter the guests of honour left the table to receive the students' tributes in the Blue Hall and were followed by the other guests. The last event of the evening was at 10.15, when the ball began. For the first time there were photographs in the newspapers of the refreshments table that was set up for the ball. It was reported that when the counter opened, the guests were already standing in line.

In 1973, the present king of Sweden, King Carl XIV Gustav presented the Nobel prizes for the first time as the monarch. In addition, the actual ceremony was changed since Sweden had a young king while the prize-winners were getting on in years. This year the King went up onto the Concert Hall platform and handed over the prizes to the laureates. This was also the year when the UN Declaration of Human Rights celebrated its 25th anniversary on December 10. Certain religious groups requested that the Nobel Foundation should take this into account when planning the festivities since they had chosen to fast to show solidarity and respect for human rights.

There were 800 guests in the Golden Hall when the Chairman of the Nobel Foundation (who had himself received a Nobel prize on a previous occasion) proposed the royal toast, after which the King called for a toast in memory of "the great donator, Alfred Nobel". Great attention had been given to the choice of music, which was performed by the Stockholm Baroque Ensemble conducted by Ulf Björlin. This was the first time a male soloist had been engaged, Sven-Bertil Taube. The arrangement was described as extraordinarily good. There were 200 serving staff and the restaurateur, Mr Ericsson, conducted them discreetly with his silver pen. The banquet concluded with the ball in the Blue Hall. None of the royal party took part in the dancing, however; they received the students' homage standing on the balcony in the Blue Hall and after the first dance, they left. The ball ended at 1 a.m.

One of the few students who were invited to sit in the Golden Hall was the Chairman of the Swedish Federation of Students Unions since he was the last speaker at the banquet. The other students sat in the Blue Hall where, like everyone else, they had place cards. Only the man's name was written, however, his partner being called, for example, "Frank's lady". But it wasn't any better in the Golden Hall, where the Swedish Prime Minister's wife was entitled "Mrs Olof Palme". The student menu consisted of gravad lax with beer and schnapps, mushroom fillet of beef au gratin with red wine and coffee with liqueurs.

If the Nobel Banquet was to have its joyful character, a considerable staff of servants was required. On the morning of the Nobel Day preparations for the evening's dishes began in the kitchen. The Nobel kitchen had been renovated, the walls were now tiled in white and it was as big as the Golden Hall. There were eight chefs this year.

The next year, 1974, the Nobel Foundation met a greater demand for tickets to the Nobel festivities and the banquet than ever before. It would have been possible to fill the whole of the Johanneshov Ice Stadium, but instead it was decided to move the banquet from the Golden Hall to the Blue Hall. There were 1,050 guests. Thus was the first year that the ladies had their own names and titles on the place cards and this was seen as a change in the times that it was hoped would have consequences. The long table, that is, the table of honour in the Blue Hall, was decorated with tall, silver candelabra and red carnations spread over the white tablecloth. At the table of honour sat 80 guests including the King and the laureates. Apart from the table of honour, there were another 50 tables decorated with candelabra, red carnations and glasses that were already filled. The hosts were described as world-class psychologists since champagne (ice-cold) was balsam even for old, radical throats. The banquet began with the royal toast, which was replied to in English with a toast in for Alfred Nobel. Karl Nilheim's Orchestra was given a spontaneous round of applause when he and his musicians led the dessert parade down the main stairway in the Blue Hall, playing the old Swedish folk march, Gärdebylåten. The ice-cream dessert consisted of 120 flame-coloured bombes glacées framed with candyfloss.When the dessert was over, student songs echoed round the brick walls of the Blue Hall. Then all conversation came to an end at the tables over coffee and liqueurs when the speeches began. Now the students ate their dinner in the Prince's Gallery. The ball was held in the Golden Hall. At 11 o'clock, the Swedish King left the festivities.

In 1975, it was time for the Nobel Foundation to celebrate its 75th anniversary. All previous Nobel laureates were invited to the Nobel Banquet, so this year there were 1,100 guests. This was also the first time that a foreign Head of State, Queen Margarethe of Denmark and her husband Prince Henrik, attended a Nobel Banquet. At 6.25, the guests took their places, at 6.55 a trumpet fanfare was sounded and the royal family arrived, and at 7.05 Professor Sune Bergström (Chairman of the Nobel Foundation, 1975–1987) proposed the royal toast. Everyone then sat down. At 7.07, the King stood up and proposed a toast in memory of Alfred Nobel, after which the dinner began. About a quarter of an hour later, the buzz of conversation in the Blue Hall rose to a peak. This banquet had 61 tables with one to two waiters per table. During the dinner, music by the Swedish composers Peterson-Berger, Karl Nielsen, Evert Taube, Bellman and Lars-Erik Larsson was played. The soloist was Sylvia Lindenstrand. The meal ended with the serving staff marching in two by two with glass serving dishes on their left hands while the orchestra played Gärdebylåten. It had to be played five or six times before all the waiters and waitresses were in position. During the dessert, the Stockholm University Singers entertained the guests with folk songs, and when coffee and liqueurs were served it was time for Olof Palme to speak for ten minutes. Each laureate had three minutes for his/her speech with the exception of the Literature prize-winner who was given five minutes. This was the first time that the Swedish Prime Minister gave a speech at a Nobel Banquet. Then followed the ball in the Golden Hall, where a refreshments buffet had been set up.

The kitchen staff had been at work since 8 o'clock in the morning preparing the food, while the tables were laid in the Blue Hall at 1 o'clock. There were 150 serving staff who came from Stockholm, Skövde, Kalmar, Göteborg and Åtvidaberg. The working day did not end for the last serving staff until 3.a.m. the following morning. Almost 250 people operated the stage setting for the banquet. This year the Literature prize-winner could not eat the food on the menu, so he was given another dish consisting of quenelles and a mild sauce. There was also another guest who asked for a peeled banana for dessert, which was provided. The staff followed a minute-by-minute schedule from 7 o'clock to 9.25, when Olof Palme's speech began.

The prize-awarding ceremony in 1976 was disturbed by a person in the audience standing up and shouting his protest against the winner of the Economics prize. When the protester came to the slogan "Crush Capitalism", the whole audience was at a loss as what to do and the King was very pale. The young protester had got in on his father's ticket and five minutes before his protest he told his mother what he was about to do. She did not believe him. He also managed to unroll a banderol before he was arrested and taken for questioning. He proudly announced that he had been practising his protest for a week.

Queen Silvia attended the banquet for the first time. Lilian Craig also made her first official appearance as a Swedish princess with the title Her Royal Highness, the Duchess of Halland. The number of guests at the banquet in the Blue Hall had now risen to 1,200, 400 of whom were students. The tables were decorated with candles and large bouquets of red and pink carnations, 1,700 in number, 70 kg of mimosa and 250 yellow chrysanthemums. The table of honour seated 80 guests and 800 other guest sat at 24 other tables in the Blue Hall. The students sat at small tables placed around the hall; in other words, they sat in the same hall as the other guests. When the glasses had been filled with champagne, the royal toast was proposed, after which the King's Song was sung and a toast was proposed in memory of Alfred Nobel. No speeches were made during the dinner, which was an innovation. The orations in the Concert Hall were thought to have been enough. Instead, the short speeches of thanks from the laureates were given at coffee. Although there was a total ban on smoking at the table of honour because the Chemistry prize-winner was strongly allergic to tobacco smoke, both Torbjörn Fälldin and Prince Bertil smoked over coffee. The dessert was carried in on trays by 140 waiters and waitresses; it was Glace Nobel with petits fours and candy floss. At their head was Karl Nilheim, the conductor of the orchestra, followed by 21 folk musicians playing Gärdebylåten on fiddles and clarinets. During the dinner, the guests listened to Nicolai Gedda and the student singers. The banquet lasted three hours and concluded with the ball.

This year the serving staff was a record 140 in number. The restaurateur, Lars Hallengren, organised his first Nobel Banquet in 1974, although that year he only looked and learned.

The 1977 Banquet was also attended by 1,200 guests and everyone had the same first course, though the main course was different for the main guests. The students had turkey, while the Nobel guests had snow grouse. The final bill for the Nobel Banquet was expensive — over 200,000 kronor.

This sum included 400 student meals at 200 kronor, which was paid by the Stockholm Student Union. Not quite half of the guests were invited by the Nobel Foundation, the remainder paid 300 kronor each. There was a financial innovation at this banquet; the guests had to pay a cloakroom fee of 1:50, but they had been warned about this on their invitations.

The seating list which the guests were given was 55 pages long. The champagne was already in the glasses when the guests found their places. Champagne was drunk with both the first course and the dessert. The other wine, 5,000 bottles at 16 kronor each, was bought already in 1975, while the champagne cost 69:25 a bottle. The students were given something described as Spanish champagne while the Nobel guests had French champagne. However, for a wine to be called champagne, it must come from the French district Champagne, so what the students got was probably a sparkling white wine from Spain.

The banquet opened with the royal toast, followed as always by the silent toast in memory of Alfred Nobel proposed by the King. The laureates' speeches were short since the dinner speeches had been abolished a few years before; it was considered that there were quite enough speeches given at the prize-giving ceremony.

130 serving staff had come to the banquet from the whole of Sweden: They had to pay for the journey to Stockholm themselves and they were paid between 250 and 350 kronor for their work, but they came for the experience of working at a Nobel Banquet. Some of them had worked at 13 previous banquets; 45 came from Uppsala and there were even some who came from Denmark. Their great moment was the march down the stairway carrying the dessert. First came Karl Nilheim and his 21-man orchestra, followed by the seven headwaiters and the 130 waiters and waitresses, marching two by two and carrying the ice-cream. The dessert was Parfait Glace Nobel, which had been introduced the year before, probably to stay. The ice-cream was decorated with a large N of gold on the top. A chemistry professor who served himself with the green candy floss observed that it was certainly not particularly healthy — but oh, how good it tasted! Three-star Martell brandy and Mandarin Liqueur Napoleon were served with the coffee. The banquet lasted three hours and concluded with the ball.

The Nobel Banquet took place on a Sunday, and on the Saturday ten chefs and four cold-buffet manageresses put out the food which they had been working on since the Monday before. On the Saturday evening,

150 litres of sauce, 150 kg of potatoes to be roasted, and 200 lettuces to be cleaned had been prepared. In addition, some 70 police officers ate in the City Hall Cellar restaurant: long, open sandwiches and fried steak and onions. The housekeeper and her staff at the City Hall had spent three days alone polishing the silver for the table decorations. This year's banquet was described as brilliant.

Foreign interest in the Nobel Banquet in 1978 was greater than ever before. For the first time, there was a major television programme about the Nobel prize-winners, and the festivities are now broadcast in the USA, Japan, Australia and South America. This year about a third of the prize-winners were of Jewish birth, a level which has remained more or less constant ever since the Second World War.

The tables were decorated with chrysanthemums, carnations and mimosa and there were covers for 1,200 guests in the Blue Hall. The schedule for the banquet was somewhat different from the previous year. It began with the royal toast and the King's Song, followed by the dinner, after which the King proposed the toast in memory of Alfred Nobel. Then came the dessert, coffee and brandy or orange liqueur, the laureates' speeches and finally the ball. The dessert was carried in to the tones of "Gånglåt från Äppelbo" played by a young group of fiddlers from Mora. The musicians were between 10 and 16 years old. The ice-cream parade was a success. The ice-cream was served with wafers and not petits fours as the previous year. When the meal was over, each laureate had five minutes for his/her speech of thanks. There were some ten guests who ate kosher food (the Physics laureate and his family) and there were also a number of vegetarians. The Literature laureate was a vegetarian and he and the other vegetarians had a menu consisting of Israeli avocado filled with low-fat curds, artichoke heart with morel sauce, vine-leaf dolmas with chopped vegetables and Nobel ice-cream. The students were given a simpler meal than the other guests: salmon canapé and moose steak, but they ate the same dessert as the others, Nobel ice-cream. For the first time, it was mentioned that the Nobel Banquet menu, in accordance with tradition, is kept secret until Nobel Day with the exception of the dessert, which after five years has become a tradition, the Nobel ice-cream. The banquet seems to have lasted four hours.

IOGT-NTO, the Swedish teetotaller organisation, demanded that the Nobel Banquet should be non-alcoholic, their main reason being that

It is time for Glace Nobel and the year is 1979. To serve the King and Queen throughout the whole of the Nobel Banquet was considered a task of honour (and still is) and you had to be very skilful indeed to be given this job. The picture shows how little space the serving staff had between the tables and between the guests, which placed great demands on the waiters' and waitresses' skills. The picture also shows the first version of the Glace Nobel that we know. The ice-cream was frozen in two-litre milk cartons. The shape of the ice-cream also reveals the need for the serving staff's skill. Each Glace Nobel should serve ten guests and not melt before the last guest was served. Nor should the ice-cream drip because then it would make the candy floss sticky and very difficult to serve. It was the same with the petits fours. © Aftonbladet bild this would strengthen Sweden's alcohol policy in the world. However, the restaurateur at the City Hall said that this would not be a good idea since alcoholic beverages were often a natural table drink for the foreign guests who attended the Nobel Banquet. However, non-alcoholic beverages were provided, as had been done for a long time.

In 1979, the tables were laid with three glasses/covers, porcelain from Gustavsberg, damask tablecloths, perfectly folded napkins, flowers and candelabra. There were 800 Nobel guests and 400 students. The price of a cover was 350 kronor and for the students 220 kronor (a simpler menu). The food matched the table arrangement. In marched 140 waiters and waitresses in good order and served the elegant entrée, to which champagne was served. A crunchy croissant was served with the hors d'oeuvre, and a red wine was served that was no longer available at the Swedish State Liquor shops. It was a 1970 vintage wine. The Nobel ice-cream was carried in accompanied by Bellman music and when this ended, all the serving staff were in position to begin serving. The ice-cream parade began at exactly 8.30. Petits fours were served with the ice-cream, followed by coffee and brandy.

The entry march itself into the Blue Hall was disturbed by a merry girl student who took the liberty of coming in unannounced before the arrival of the royal party and standing on the staircase, where she sang that people should love each other more. She was gently led back to her seat but when she did the same thing again at her table, she was taken out by the security guards. The speeches by the laureates were so short that two of them decided to do theirs together.

The retiring restaurateur revealed that he had faced a difficult problem when one of the laureates had no teeth and so could not chew his food. The problem was solved by grinding the meat, cutting the fish into thin slices and mashing the berries. He also pointed out that the Nobel Banquet is a precision job whose planning begins in September. That is when the Nobel Foundation requests tenders for the banquet. When a tender has been accepted, the planning begins: what is the menu to be, three courses with a first course, a main course and a dessert. The food should be as Scandinavian as possible. Salmon, reindeer, moose, venison or grouse for the main course were desirable, and cloudberry, lingonberry and blackcurrant sorbet for the dessert. In October, the Nobel Foundation with their ladies eat the menus, which are criticised and altered. After that, the dinner is decided on at the same time as the invitations are sent out. Serving the dinner is a race against time, with a margin of error of at most 1½ minutes. The speeches have to be short, so all the laureates have 3½ minutes except for the Literature prize-winner, who has 5½ minutes at his/her disposal. The festivities reach their climax when the Blue Hall is darkened and the dessert is slowly carried in a procession lit up by floodlights down the stairway, a ceremony that many people never forget. The royal couple are always served by two specially chosen waiters, while elsewhere each waiter has ten guests to serve.

The 1,200 tickets to the banquet in 1980 had been taken up long ago. The Nobel Foundation was already swamped with requests from people who were queuing up for the next year's celebrations. This was the first year when brandy or cognac were not served with the coffee after insistent requests from the teetotallers associations. The guests would have to make do with champagne, red wine and sherry. Queen Silvia could not attend the banquet because she was ill.

The cost per cover this year was 350 kronor. 3,600 plates and 6,000 pieces of cutlery were needed. The tables began to be laid at 1 o'clock and were ready at 4.30. One table of honour with 80 places and 24 side tables with 40 places each, plus 40 tables with 10 places each in each of the Blue Hall's arcades had been laid. Damask clothes on the tables, Gustavsberg porcelain, three kinds of glasses, the starched napkins folded artistically together with the candelabra and the flower arrangements made a beautiful setting. The last things to be put out were the place cards. An innovation this year was that the Nobel Foundation had ordered a type of candle for the candelabra that did not run, since there had been problems with candle wax ruining the tablecloths and dishes.

The guests were requested to take their places in the Blue Hall at 6.35. At 7 o'clock the King and Queen and the laureates with their families arrived. Just before this, the restaurateur signalled to the orchestra, which played a fanfare and the whole assembly stood up. Then followed the usual banquet ceremonies. The Nobel ice-cream this year was crowned with an N in gilded croquembouche. The laureates gave their speeches of thanks lasting three minutes each.

The kitchen and serving staff comprised 6 headwaiters, ten or so chefs and cold-buffet manageresses and 130 waiters and waitresses — all professionally qualified. The serving staff together walked 600 km during the evening. Serving at the Nobel Banquet was considered an attractive job but has also been called the waiters' and waitresses' Vasa ski race. The oldest waiter was 77 years old and was working at his 31st banquet.

All the courses were eaten on time. The year before, however, the King had had such a charming lady at the table that he ate too slowly. In addition, the press photographers devoted too much time to Queen Silvia, which delayed the meal by a whole ten minutes. It is seldom that not everything goes completely smoothly at the dinner, but the human factor can of course play a role here too. One year when the Literature prize-winner was a vegetarian and was going to have a different menu consisting of vegetables, there was a mix-up. He only poked at a lettuce leaf and ate some bread, leaving the salmon, while another guest, to his great surprise, got a completely vegetarian hors d'oeuvre.

The Banquets, 1981–1990

In 1981, the City Hall succeeded in squeezing in another 50 guests at the banquet, so that the total number was 1,250 persons, including 400 students. This was a completely reversed situation compared with the 1901 banquet, when the Nobel Foundation was obliged to call in people by appealing to government departments and companies. The cover price in 1901 was 15 kronor for a five-course meal, and 375 kronor for this year's banquet. The students paid 200 kronor but not for the same food as the other Nobel guests, only a simple first course — though not all that simple. How about a canapé of whitefish roe and shrimps? This year was the 80th anniversary of the Nobel prizes.

At 1 o'clock, they began to lay the tables in the Blue Hall, using a measuring stick to measure the distance between the tables at exactly 1.45 metres. The distance between plates was measured with the help of shorter sticks. The restaurateur directed this work from the marble steps with a thunderous voice by means of a cordless microphone. The waiter who had served at most Nobel Banquets worked for the first time at the 1944 banquet. For the serving staff the Nobel Day began at 3.30 and at 4.30 they all met for dinner, when the head-waiter called the roll and made sure that all those who had applied had actually come. He then handed out the order of the day, which set out in detail everything about the food, the drinks and the times. There were ten guests each waiter/waitress had to look after at the main table while at the student tables they had to look after 15 guests. Thus the service at the table of honour was faster. After having gone through the order of the day, the staff gathered in the Blue Hall where all the waiters/waitresses took up their positions to be checked again. Serving the food in the evening was controlled by signals from the first headwaiter, and woe betide the person who served too early! This veteran headwaiter often looked after newcomers who were often very nervous, which is the worst thing to be. The worst accident he had experienced at the banquet was when a waiter tripped up on a chair leg and fell flat on his face. However, he managed to catch himself with his elbow so that the food on the plate survived!

This is the first time it was mentioned that the journalists covering the banquet had to sit at a special table of their own, the reason being that they had proved to be bad table partners. Their conversation and witty remarks were restricted to what they wrote down in their notebooks, so they all had to sit together.

The kitchen staff consisted of 12 chefs and cold-buffet manageresses, and there were 130 waiters and waitresses. Those who served the King and the Queen were both women; the waitress who looked after the King was working at her 26th Nobel banquet and the waitress who served the Queen was working at her 10th Nobel banquet. Serving at the table of honour was seen as an honorary task. When the restaurateur gave the signal for the champagne and wine to be uncorked after it had been tasted, ten openers went to work under the supervision of the wine master. These ten people opened 450 bottles of champagne, Mumm Cordon Rouge for the Nobel guests and Codorniu for the students, and 325 bottles of red wine, Chateau St Germain for the Nobel guests and Paternina Banda Azul for the students. The wine is aired for at least three hours before being served. The wine master said that he himself with two assistants opened 100 bottles in a quarter of an hour using a double corkscrew, a corkscrew with two opposing screws. An effective cork-killer for amateurs but Sigge and his assistants were masters of their instruments. The housekeeper had ten dishwashers to help with the washing-up.

The Literature prize-winner was a diabetic and was given the menu he asked for and non-alcoholic drinks. His menu was vegetarian, consisting of avocado and hot vegetables. The dinner took three hours and began with the King proposing the toast for Alfred Nobel. There were 140 dishes of ice-cream and the laureates' grandchildren were wide-eyed and the

Queen applauded when the dessert was carried in. The student songs were performed over coffee. The banquet concluded with the ball in the Golden Hall.

While the banquet continued in the Blue Hall, the Golden Hall was used to line up the round, silver dishes that the ice-cream would be served on. These dishes were decorated with an N in coloured glazing. The Nobel ice-cream was carried in to the tones of folk music played on fiddles. As well as being the grand finale of the banquet, it was also the dread of the serving staff. The ice-cream was high, hard to cut up and slippery on the dish. It was served with green candy floss, petits fours and wafers. One of the laureates this year was helped by the Queen to fold his programme into an envelope in which he put the coveted N from the bottom of the ice-cream. The ice-cream was now described as obligatory. The brandy and liqueurs with the coffee were absent but at the ball a refreshments table was set up where the guests could buy subsidised glasses of whisky while the Nobel Foundation stood for the soft drinks. After the banquet, however, the Nobel prize-winners, the ambassadors and the members of the government disappeared upstairs to the Gallery, where free drinks were served and the Press were strictly forbidden.

The head chef described the menu as 100 per cent Swedish without any complicated dishes. An innovation was the Nobel roll of coarse rye served without butter with the first course. The salmon mousse was mixed the day before, covered with lobster and mushroom. It was described as having a distinct, delicate salmon flavour with a consistency as light as a butterfly. The sauce verte was cold and consisted of mayonnaise spiced with chervil, tarragon, cress, chive and spinach. The moose meat was marinated for three days in red wine, pepper, onion, juniper berry and thyme, resulting in a very tender and tasty dish which was accompanied by a phenomenal morel sauce painstakingly cooked with finely chopped onion and whipping cream. The garnishings were rowanberry jelly, brussels sprouts and potato poireau fried with shredded leek. This sauce was described as a miracle as 100 litres of it were cooked but it tasted as it if were made for four persons.

During the banquet in the City Hall, many of the staff were victims of a theft; someone broke into the staff cloakroom and took clothes, money and the like, but it was not known exactly how much was stolen. Otherwise this banquet was described as the best ever.

Interest in attending the 1982 Banquet was unbelievable. When only a week remained, there were 20 guests who had to be included for whom there was no room. The only hope was that there would be some cancellations. The Nobel Banquet in the Blue Hall was described as lightest, happiest and liveliest for many years. There were 1,310 guests, 400 of whom were students.

At 1 o'clock, 25 people began to lay the tables with newly starched damask cloths on the head table, 27 side tables and 40 tables in the arcades, freshly folded napkins, the gilt-edged porcelain from Gustavsberg, cutlery and three kinds of hand-blown glasses from Älghult. The only food that was served at this time was meatballs for the police sniffer dogs who worked their way through the City Hall. At 4 o'clock, a coach arrived with 40 waiters and waitresses from Uppsala. At 4.30, the serving staff gathered for dinner in the City Hall cafeteria. There was a roll call, detailed information was given and the boss gave a pep talk like the ones given before an ice-hockey match. Everyone was given an order of the day showing them their duties minute by minute.

At 5.30, the staff placed out bread rolls and water; the water was poured into the glasses on the head table. Champagne was served at all the tables except at the main table, where it was not served until everyone had arrived. At 6.30 the guests began to seat themselves, at 7 o'clock the King and the Queen arrived together with the laureates and their families. By then there had been a fanfare of trumpets, the guests had stood up and when everyone was in place, the Deputy Chairman of the Nobel Foundation proposed the royal toast. This toast was normally proposed by the Chairman of the Nobel Foundation, but this year he was otherwise engaged as one of the laureates. The King then proposed the toast in memory of Alfred Nobel. The ice-cream parade was more festive than usual thanks to 60 Colombian dancers and singers who led the parade — this was in tribute to the Latin-American Literature prize-winner. There were seven speeches after the dessert, and with the coffee came the student choir. At 10 o'clock, when the ball began, the kitchen staff sat down and ate the Nobel dinner.

The guests could converse without difficulty across the table without any decorations in the way. The flowers, which were a gift from San Remo, were arranged in low bouquets.

The banquet cost about 400,000 kronor, but most of the guests paid 400 kronor themselves and a few researchers got a 200-kronor discount. The food bill for the evening was 50,000 kronor for the Nobel Foundation. The dinner had grown out of the Blue Hall, so some of the guests, including the 20-man strong corps of journalists that the Literature prize-winner had taken with him, had to sit in the galleries, where all the police officers, the secret servicemen, the student singers and the attendants also ate. The banquet had become world famous for its perfect service down to the slightest detail and absolutely first-class food, which was matched by wine served at exactly the right temperature. Apart from the waiters and waitresses, there were eight headwaiters, twelve chefs and cold-buffet manageresses, ten dish-washers and ten bottle-openers employed for the banquet.

The 1983 Nobel Banquet was also the 150th anniversary of the birth of Alfred Nobel. Specially invited guests were to arrive from abroad, but the number of laureates was smaller than usual. Six books had been published to celebrate the 150th anniversary of Nobel's birth, so one whole table at the banquet was reserved for authors and publishers. The number of guests was set at 1,100 plus 150 students and a ticket cost 500 kronor, yet there was a long waiting list for the banquet. After the ice-cream parade there would be student tributes and singing for the laureates.

At 1 o'clock the serving staff were at their posts after the attendants in the Blue Hall had measured that the 66 tables were in the right positions. The table of honour seated 78 guests. The tables were then laid and the distance between each cover was measured with measuring sticks.

The banquet began with the Pomp and Circumstance March while the royal party and the whole troop of Nobel officials and laureates marched down the steps from the Golden Hall to the Blue Hall. The royal toast in Pommery champagne was proposed by Professor Sune Bergström. Two minutes later, the King proposed the toast to Alfred Nobel. The traditional students' speech to the laureates and a laureate's response had been taken away from the thick programme. Instead, the tribute consisted of song and music. The ice-cream was carried in at 8.28, two minutes earlier than the timetable. The Mora Youth Folk Band led the traditional ice-cream parade, which consisted of 120 waiters and waitresses. The Glace Nobel was crowned with the letter N in gilded croquembouche. Prince Bertil's chef was there to make sure that the N was put in the right place on the ice-cream. The dinner lasted almost three hours, and after the dessert came the laureates' speeches of thanks. That was when the sniffer dogs were eating their meatballs and cold roast beef. The last member of the staff to leave the City hall was the housekeeper, who went home at 4 a.m. Leif Kronlund's Big Band played at the ball, which is said to have gone on until 6 a.m.

The head chef had created the menu for the Nobel guests by first composing four different menus that were sent to the Nobel Foundation. The Foundation picked out the most interesting parts and then contacted the City Hall to arrange a trial dinner. The decisive factors when planning the menu were that it should be rational to serve and that it should be good in taste and appearance.

Another important factor when putting the menu together was that it should have a Swedish character for the sake of the foreign guests, and that the whole festival should be something of a national Swedish manifestation; on the other hand, the food should not be too unusual and exotic for the foreign guests. Pickled herring, fermented herring, potato dumplings stuffed with pork or even gravad lax would definitely not fit this context. Salmon, however, occurs in many different forms on the Nobel menus. What is important is the colour, consistency and taste of the dishes; furthermore, the year's Nobel banquet must not be like the previous year's , nor be identical with any of the other previous dinners.

Preparing the food was yet another decisive factor since it had to be made for 1,300 people. Admittedly, the kitchen was spacious, yet it had a limited capacity in the number of ovens and hotplates. From a serving point of view, it was also quite impossible to serve soufflés or food portions. There were no tables long enough, and none of the cold-buffet staff had arms long enough to fill and garnish 130 hors d'oeuvres at one time. The starters could be served as portions on the plate but then they would lie on a lettuce leaf on the serving dish to be served to the guests from that dish. This year, the head chef had brought in five extra colleagues apart from the ordinary staff of six in the kitchen.

Since the banquet was held in the Blue Hall instead of in the Golden Hall, the chef thought that they had got an excellent serving room before and during the dinner in the Golden Hall. As soon as the ice-cream was served, all the tables were cleared and moved out of the Hall so it was as empty and shining clean as it was before the banquet began, ready for the guests to enter the dance floor.

Altogether there were 220 persons who were responsible for the guests' festivities, including 12 cooks and cold-buffet manageresses, 10 bottle openers, 15 people at the disposal of the housekeepers, 140 waiters and waitresses, 6 headwaiters and many, many more. The guests drank 260 litres of coffee, 600 bottles of Pommery & Greno, 200 bottles of Château Montfaucon and 1,400 export-type bottles of Ramlösa mineral water. These bottles were more elegant than those used at ordinary conferences. If one waiter was to serve all the guests, it would take him 780 hours or 32 days and nights, but luckily each waiter/waitress had ten guests to serve, no fewer, no less.

The 1984 Banquet was a splendid display of the Swedish ability to organise large parties within carefully prepared limits. The table of honour seated 79 guests: the laureates, the royal party, representatives of the government and eight foreign ambassadors. Two of the guest of honour did not attend; one was Prince Bertil, who had injured a rib, and the other was Tord Magnusson, the husband of Princess Christina, who had a cold. The total number of guests was 1,350 and there were 250 students. The cover price was 550 kronor. The 25 tables were decorated with candelabra, 10,000 carnations and 10 kg of mimosa.

A young Swedish girl student was given the honour for the first time of being the toastmaster, replacing the legendary Dr. Wilhelm Odelberg. The choice of toastmaster was considered very bold by the Nobel Foundation since previously this role had been held only by men. Dr Odelberg had played the part for 20 years before the young girl student took over. The toastmaster's duty was to announce the arrival of the royal party, the toasts and the speakers. The speakers were to be introduced in their own language. The young student toastmaster took part in the previous year's banquet as a standard bearer.

The entry of the many guests of honour was accompanied by music from the opera Tannhäuser, and during the dinner a group from the Stockholm Philharmonic Orchestra entertained the guests, while the dance music in the Golden Hall was played by the best-known musicians on the Swedish jazz scene, Arne Domnérus Big Band with Putte Wickman and Bengt Hallberg as soloists.

The royal toast was proposed by the Chairman of the Nobel Foundation. The dinner was described as glorious, with an extremely delicious menu. Table 12 at the banquet was as usual reserved as "the children's table", where all the children and grandchildren of the laureates sat and drank each other's toast in sparkling Pommac.

The banquet was described as a gigantic family party in the Blue Hall, where everything ran according to plan. At exactly five past seven, the King proposed the toast in champagne for Alfred Nobel. As soon as the formal ceremonies with the official toasts were over, conversation in many languages echoed from the Blue Hall's walls and ceiling. But when the ice-cream parade, a whole army of waiters and waitresses dressed in black and white, marched down the stairway, there was instant silence. The folk-music band from Skansen launched into Kers-Lars bridal march from Leksand and the whole ice-cream parade set off down the stairs. An innovation here was that the traditional blackcurrant sorbet was replaced by kiwi sorbet covered with passion-fruit sorbet.

As usual, the Glace Nobel was decorated with a large N and Queen Silvia made sure that her dinner partner and laureate took an N home with him as a souvenir. The dinner lasted three hours. The student's entertainment was performed by a male choir.

One of the requirements made of the serving staff was that they should be able to carry 15 plates at a time. The 145-man strong staff comprised all ages, from students to pensioners. Covers A19 (the King) and A59 (the Queen) were served by pensioners. The waiters and waitresses came from all over Sweden, some of then working for nothing, others even losing out on the deal since they paid for their travel and accommodation themselves. At 12.30 on Nobel Day, 25 staff arrived to start laying the tables; with the aid of a two-metre-long measuring stick, they made sure that the distance between the covers was identical. At 4.30, the rest of the serving staff arrived and they all had dinner together before running through the programme. During the banquet, the manager stood on the Blue Hall stairs, directing his staff with discreet signals — a nod meant that the King and Queen should be served, and when he lowered his hands it meant that the other guests should be served. The flower arrangements were now the only table decorations. Previously, the tables had also been decorated with mirrors and large, heavy, gilded candelabra with candles and flowers on them, but now the mirrors were absent and the large candelabra had been replaced by smaller ones. The housekeeper had 17 persons to help her, and the other staff comprised eight headwaiters, ten chefs and cold-buffet manageresses and four wine waiters.

There were 1,280 guests at the 1985 Banquet. It cost the Nobel Foundation about 300,000 kronor, but those who were not specially invited had to pay 500 kronor for a ticket. At 12.30, the serving staff gathered in the City Hall, and by then 15 chefs and cold-buffet manageresses were already working hard in the kitchen. At 12.45, the staff began to set up the tables and chairs in perfect rows with the help of a measuring stick, and 25 minutes later all 66 tables were covered with cloths and the porcelain was on its way in on large trolleys. Laying the tables was a science of its own in which everything should be accurately placed down to the last millimetre. The distance between the head table where the royal party and the laureates sat and the tables alongside it had to be exactly the same. All the plates had to be exactly opposite each other, because if they were not, it would look as though the tables went in waves. A special waiter measured the distance between plate and glasses. The table decoration at the head table was a gift from a German family who had presented it when the City Hall was inaugurated. It consisted of glass and gilded metal. The 4,000 red carnations and 10 kg of mimosa which formed the flower arrangements on the tables were a gift from the Mayor of San Remo. The previous year's toastmaster had done such a good job that she was entrusted with the duties this year too.

The banquet had the following timetable: at 6.45, introductions in the Prince's Gallery and the guests were requested to take their places in the Blue Hall, wine was served at the head table; at 7.05, His Majesty the King's toast was proposed by Professor Sune Bergström; at 7.07, the toast in memory of Alfred Nobel was proposed by the King; at 7.10, a two-minute photoshoot of the King and Queen and five minutes for the other guests; at 8.30, dessert served. The head chef himself took part for the first time in the ice-cream parade, which was led by the Floda Folk Music Band, closely followed by the head chef who was over six feet tall plus his high chef's cap, and he in turn was followed by the waiters and waitresses. In the kitchen the staff were fed up with all the guests who came in to get a bag for their caramel souvenir, so they had got together a pile of plastic bags and paper serviettes to give to the people who had managed to get hold of a caramel N. At 8.55 coffee was served, with a fanfare before each laureate's speech. Brandy was not served with the coffee, it had been cut out of the menu, which saved ten minutes in the timetable. At 9.10, the speech by the Literature prize-winner, at 9.15, by the Medicine prize-winner; at 9.21 by the Chemistry prize-winner; at 9.22 by the Physics prize-winner; at 9.25 by the Economics prize-winner and at 9.30 the students' tribute in the form of songs by the Stockholm Student Choir led by Lars Bohm. At 10 o'clock, the Ball commenced immediately after the student choir had finished.

The first parade was when the guests of honour entered the Blue Hall at the beginning of the banquet, marching down the stairs to the tones of the Pomp and Circumstance March. The soloist during the dinner was Lill Lindfors and the Stockholm Student Choir performed the tribute to the laureates. The banquet concluded with the ball at which Leif Kronlund's Orchestra played and Lill Lindfors sang.

The menu was described as excellent. Moët-Chandon champagne was served with the first course and a specially imported 1979 vintage wine, Médoc Baron Philippe, with the main course. At 1 a.m. the banquet was over, and half an hour later it was impossible to see that there had been a Nobel Banquet in the Hall.

The ice-cream was the most spectacular part of the dinner. It was delivered to the City Hall a couple of hours before the banquet started, and it was a wonderful sight to see it lined up in the Golden Hall before it was carried down to the guests below. The Golden Hall also acted as a dish-washing room before the ball began, probably the most beautiful dish-washing room in the world.

The head chef stated that the food had to be hardy (a suitable description for a meal that is prepared en masse), but so tasty and attractive that the guests would remember it. He had ten chefs and cold-buffet manageresses and five extra assistants at his disposal (many people worked in the kitchen for nothing, just to take part in a Nobel banquet), in addition to whom there were ten bottle-openers, fifteen dish-washers and 140 waiters and waitresses. When all the work was done, the police officers, the stewards, the artists and all the others involved had their own Nobel dinner in peace and quiet.

The reason for keeping the menu secret was that otherwise all the restaurants would advertise: "Come and eat the Nobel dinner with us" on Nobel Day. What was new this year was the massive coverage of the kitchen and its staff by television. As early as 7 a.m. television set up its cameras, and what went on behind the scene was covered more closely than even before.

The menu had to be first-class and it ought to consist of Swedish-style food. By late summer the menu proposals had crystallised. During the autumn, the Nobel Foundation test-ate the proposed dishes. The dishes that were rejected were: 1) Bohus canapé (crab and shrimp in leek mayonnaise), duckling in orange sauce, buttered potatoes with parsley and fresh green salad; 2) crayfish canapé with basil sauce, Nobel roll, saddle of lamb garni with red-wine sauce and matchstick potatoes fried raw; 3) cold pheasant in gravy with apple and mandarin salad, Nobel roll, smoked salmon with mousseline sauce, marinated asparagus with potatoes boiled in dill; and 4) eel-stuffed turbot timbale and green sauce, Nobel roll, fillet of venison with celery cooked in cream, lingonberry-cream sauce and buttered potatoes. Glace Nobel was the same for all the alternatives, that is, passion fruit and kiwi. The students who attended the banquet had to make do with a selection from the menus that were rejected for this year's menu.

The banquet also included some of the people who were working in the periphery: 60 police officers, 25 musicians, 80 radio and television employees and 10 people from the Security Police and the Press.

The rumours about the Nobel Banquet being so splendid led big international companies like ITT, Ford and others to come to Stockholm after the banquet and order exact copies of the whole Nobel menu.

The 1986 Nobel Banquet differed from the previous years' in that the security check was meticulous as a result of the murder of Olof Palme nine months before. None of the members of the academic community were let into either the prize-awarding ceremony in the Concert Hall or the banquet in the Blue Hall without showing their ticket and ID.

Nonetheless, this year's festivities were described as magnificent. There were 1,250 guests at the banquet. A few minutes after 7 o'clock, the 140 serving staff marched in with the first course at a signal from the restaurateur. The laureates and the royal party had just seated themselves at the table of honour which, like all the other tables, was decorated with red carnations and yellow mimosa, a present from San Remo. The King and the Queen were served first as was only right and proper. The waitress who served the King wine had worked at the Nobel Banquets for 26 years and she and her colleagues had, over the years, walked nearly the equivalent of the circumference of the earth in their trips between the Golden Hall one flight up that acted as the serving room and the Blue Hall. These waiters and waitresses knew the royal guests, that, for example, the Queen only drank white wine because red wine gave her a headache, that Princess Lilian could not eat lobster and that both the

King and the Queen wanted small helpings so that they would be able to put their knives and forks down at the right time and not upset the very exact timetable. The clever banquet manager had placed the guest so that those who might have the same interests were placed together, hence the table for young people, Number 44 of the 66 tables. Although these young people were in the worst hamburger age, the food was given top marks. Outside the programme, a toast was proposed for Prince Bertil, who had taken part in the Nobel banquets for 50 years.

The ice-cream parade was not served to the accompaniment of folk music but with a group of five jolly girls who danced the can-can with their skirts over their heads, to the guests' great delight. The Director of the Nobel Foundation was a little worried that some people might be shocked by the can-can girls, but he was proved wrong. The ice-cream was decorated with petits fours, candy floss and an N as in Nobel made of caramel. The candy floss round the ice-cream was only for decoration but would be served to the guests if they expressly asked for it. This year as before the guests fought for the coveted Ns on top of the ice-cream. The ice-cream was served this year in slices instead of the traditional slabs. After the dessert, the laureates gave their three-minute speeches of thanks, except for the Literature prize-winner, who spoke for five minutes. It was then time for coffee before the meal was succeeded by dancing and refreshments to the music of Leif Kronlund's orchestra. The dinner lasted almost three hours, one hour of which was devoted to speeches and the students' tribute. The price per person was 600 kronor, everything included (food, drinks, service, cloakroom etc.). Some of the guests, however, paid the subsidised price of 275 kronor.

The Literature prize-winner, who had a sense of drama, was fascinated by the banquet's strict choreography; even the waiters and waitresses moved in fixed patterns. The royal toast was proposed at exactly 7.05 and the King's response two minutes later. The banquet was described as a colourful feast. There was more than the usual secrecy about the dinner, the suspense had to be maintained for the sake of the guests, as the head chef put it. If the menu was not kept secret, other restaurants would copy it and serve the same food on Nobel Day, so the mystery was important. Four menus were presented in September, two were selected and test-eaten a few days later by representatives of the Nobel Foundation. Despite all the secrecy, information about the first course leaked out and was revealed two days before the Nobel Banquet. The head chef had only one comment: Gangsters!

Kitchen staff queued up to work in the kitchen for a few hours without pay. One of the cold-buffet manageresses working alone made the 1,300 spinach beds for the salmon, which was the hors d'oeuvre, and one chef stood for hours frying potatoes; in spite of the sometimes monotonous tasks, the atmosphere in the kitchen was good. Many of the staff returned year after year, travelling a long way and meeting only there. That was what gave the kitchen that special feeling of camaraderie; the jokes and the tone were what might be called hearty. There were no specially marked out serving dishes that were destined for the table of honour. The aim was that all the dishes should be equally beautifully arranged. Working in the Nobel kitchen was a job that many felt themselves called to, but few were chosen by the head chef, who had to say no to about 20 chefs. The banquet manager had to say no to 150 voluntary waiters and waitresses. A sixty-year-old waitress did her thirtieth Nobel Banquet in a row this year.

Many who wanted to serve put their names on a waiting list to get a chance to work at the banquet, but the list was usually full several months in advance. People came from all over Sweden. Half past four was the time when the tables in the Blue Hall should be laid. Five hours earlier the Hall had been empty and deserted and then an efficient working force of 25 men came in and set up all the tables. When this was done, the ladies came in, spread the white linen tablecloths and put out the plates, the cutlery and the three glasses, one for the champagne, one for the red wine with the main course and the third for the port that was served to the ice-cream. 1,300 white linen napkins were folded into fans and placed on the plates, the same number of royal-blue menu cards were set out. In the middle of the table of honour were laid gilt-edged mirrors on which the flower arrangements and candelabra were placed. The distance between the plates was measured with measuring sticks.

About 1,400 bottles of mineral water, 260 litres of coffee, 300 bottles of champagne, over 5,000 plates and more than 9,000 pieces of cutlery were used. Apart from the kitchen and serving staff there were about 60 police officers, 25 attendants, 80 people from radio, television and the Press and an unknown number of security guards.

In 1987, there were 1,200 guests including 250 students and the price per person was 700 kronor. Half of the guests were women. This time, people from both Sweden and abroad were on the waiting list to get a ticket. The day after the Nobel Banquet, the remains of the dinner were served as the dish of the day at the City Hall restaurant for 36 kronor a plate. The menu was secret, but what could be revealed in advance was that the main course would not be venison because that was what was usually on the King's menu on the 11th December. 320 bottles of champagne, 4,200 red carnations and eight kilos of scented mimosa, lit candelabra and 7,800 pieces of silver cutlery were needed for the Nobel Banquet. There were 140 tables and 145 waiters and waitresses.

A change in the music accompanying the entry of the guests into the Blue Hall was that the Pomp and Circumstance March was replaced by the Karelia Suite by Jean Sibelius. King Carl Gustaf proposed for the sixteenth time a toast in memory of Alfred Nobel at the beginning of the banquet in the Blue Hall and Arja Saijonmaa sang during the dinner.

The food was carefully chosen and composed. While the menu was being enjoyed, the atmosphere lightened and the dinner turned into a feast. Around 11 o'clock, the Medicine laureate declared that the festivities had now begun and he disappeared rapidly out onto the dance floor. The royal party left at 11.30. Chalmers Alliance Orchestra had been engaged to play fun and games for ten minutes this year.

The ice-cream parade passed off without incident. The Glace Nobel was blackcurrant parfait filled with elderberry flower sorbet and passion fruit sorbet, petits fours, almond wafers and candy floss, with a shining green N on top. Oscar Romberg, the waiter who served the dessert to Queen Silvia, was welcomed back by the Queen. Oscar had served her for ten years but had been ill the previous year. A big question seemed to be which guests had succeeded in being the happy owners of a Nobel N in caramel that crowned each dish of ice-cream parfait. Most of these Ns, according to reports, used to be sneaked into secret vanity bags to reappear in the homes of certain Östermalm ladies at their Christmas lunches. The ice-cream parfaits were lower than usual and built in two round storeys. In previous years they had been frozen in rectangular milk packets.

Since 1965, a coach from Uppsala had taken some 30 people to Stockholm to serve at the Nobel banquet, and 1988 was no exception. Hardly any of then were full-time waiters or waitresses, they only worked as temporary staff, yet it was just these people who were hand-picked for the banquet. Before the banquet began, the serving staff had dinner, fillet of pork and cod in shrimp sauce, at Restaurang Eken in the City Hall.

The atmosphere at this dinner was very good since they all knew each other. The banquet manager handed out the timetable so that everyone knew whom they would be serving.

Those serving the students had a more stressful job because they had 15 guests to look after while the others only had ten. The person who had the honour of serving the King was doing so for the eighth time. At last, the students now had the same menu as the other guests, but they still had different drinks. The serving staff were divided into three groups: table-laying, serving and table-clearing. The staff from Uppsala were all in the serving group. The tables were laid by 4.30 and it had taken four hours to get everything in place, including the 2,500 carnations and the mimosa from San Remo.

When the guests had arrived, the serving staff stood at attention with their right hands behind their backs, waiting for the banquet manager's signal to start serving. A glance was enough to tell the table of honour that it was time to serve the royal couple, after which a sweeping gesture told the rest of the staff that they could start serving the other guests. It was the banquet manager and the eight headwaiters who directed the banquet.

There were 66 tables, decorated with red carnations, mimosa and green foliage. It needed 450 metres of linen tablecloths, 5,200 plates and the same number of glasses and pieces of silver cutlery to set the table. The were 1,058 guests and so many of them were children that they filled a table of their own. The banquet began with the royal toast and the King responded with the toast in memory of Alfred Nobel. The ice-cream, lingonberry parfait and apple sorbet, looked like pink igloos with a large, golden caramel N on top; it was served with port wine. For the first time, the ice-cream parade was accompanied by the Bleckhornen brass band. The speeches of thanks by the laureates came after the dessert, but there was no Literature speech this year as the laureate himself was not in Stockholm; instead, he was represented by his two daughters and his speech was replaced by music from his home country, Egypt.

The man in charge of the music was Arne "Dompan" Domnerus with his ten-man band and Sissi Sjöblom as the vocalist. The students were represented by Bleckhornen and the Stockholm Student Choir. Åsa Jinder was the evening's soloist, playing the Swedish nyckelharpa.

The banquet seems to have been held in a mood of warmth as laughter and high spirits hung in the air and it was not as formal as it had been the year before. In addition, the ice-cream was even tastier. At one of the tables they discussed how the serving staff could possibly manage to bring the food to the table while it was still hot, because the kitchen lay miles away from the Blue Hall. A lady who was something of a regular at the banquet revealed that she had abandoned her vegetarian diet at this dinner so as not to make life even more difficult for the waiters and waitresses. The banquet was described as a feast full of smiles and laughter.

The menus that were not accepted by the Nobel Foundation this year were: 1) Pheasant mousse, salmon schnitzel filled with duck liver; 2) Salmon and turbot in champagne jelly, minced moose meat; 3) Eel and turbot terrine, breast of black grouse. The dessert on all the menus was the classic Glace Nobel, but the garnishing varied from year to year and this year it was Swedish wild berries and fruit.

Of course, it is fun putting together a Nobel banquet but at the same time difficult to vary it since the raw materials should be Swedish and it takes months to get hold of them. It was not possible for the restaurateur to pick out one particular favourite menu from over the years because, as he pointed out, the one he was working on at the moment was always the best. What was considered good banquet food had varied over the years, however. In the 1950s, it was chicken, which appeared on all festive occasions, whereas in the 1980s and 1990s it was saddles of meat such as reindeer, moose or deer, fish like salmon, Dover sole and turbot and game birds of various kinds like pheasant and wild duck that were considered festive food.

The 1989 Banquet was attended by 1,308 guests and cost 800 kronor per cover. It needed 22,100 household utensils to set up the banquet.

At 7.05, the Chairman of the Nobel Foundation proposed the royal toast to which the King responded two minutes later with the toast in memory of Alfred Nobel. This year's Literature prize-winner, Camilo José Cela, had written at some time "Wine does a great deal for a woman, especially if it is drunk by a man", and he enjoyed the banquet's excellent wines, especially the 1982 Médoc Nobel Selection from Baron Rothschild's famous vineyard. After the main course, this year's other main literary figure, Boris Pasternack, was paid a posthumous tribute, 31 years after he was not allowed to come to Stockholm to receive his prize; instead, his son had to take it on his behalf. This was the first time that a foreign orchestra, the Oslo University Student Orchestra, led the ice-cream parade. 150 Nobel Ns were needed for the ice-cream, which the guests took home with them, along with their place cards and menus. The ice-cream was decorated with candy floss and petits fours. After the dessert, the laureates gave their speeches of thanks. On the stroke of ten, the guests rose from the table and went to the ball in the Golden Hall, where Leif Kronlund's orchestra played.

This year, the City Hall had a new restaurateur who also had his debut at the Nobel Banquet, which he managed with distinction. Everyone was impressed by his achievement. The Literature prize-winner had brought with him observers from his favourite restaurant, Los Remos, and they were very impressed by the back-stage arrangements. However, the Spanish chefs cut the meat slices too thick, so there was nearly not enough to go round.

On the other hand, there had never been so much ice-cream left over after the guests were served. The Minister of Finance at that time did not have time to eat his main course before the table was cleared, but he said it was his own fault because he had had such a very nice conversation with his dinner partner instead of eating. During the rest of the year, the City Hall restaurant sold 10–12,000 Nobel menus.

The cover price in 1990 was 900 kronor. There were no reservations or requests from the laureates concerning the menu this year. It was a pleasure to serve the King and the Queen because they liked almost everything. The programme followed the traditional, well-tested pattern. The Chairman of the Nobel Foundation, Lars Gyllensten, proposed the royal toast and the King responded with the toast in memory of Alfred Nobel after which the hungry guests could begin to eat.

The food was described as delicious. After the main course, trombone music was played and then the dessert was served. The climax of the evening, the ice-cream parade, was led by the Royal Institute of Technology's Orchestra, which played "Tomtarnas Vaktparad" (The Parade of the Father Christmases). The ice-cream still had its candy floss as decoration although it had been requested that it should be taken away in view of the fact that there were so many laureates who had reached such a venerable age that they had no teeth left. After the dessert came the laureates' speeches of thanks, followed by the student's tributes. It was now time for the ball to the music of Anders Berglund's 23-man orchestra with Melissa Murray as vocalist. The serving staff included three guest catering-school students.

The Banquets, 1991–2001

In 1991, the 90th anniversary of the award of the Nobel prizes was celebrated. The Stockholm Globe Arena was the venue of the prize-awarding ceremony since the Concert Hall was now too small. A new table setting was created for this anniversary in the form of a twelve-part set of Gustavsberg bone porcelain with a clear glaze and gold edges designed by Karin Björquist. Gunnar Cyrén at Orrefors designed the five glasses, the classic carafe and the cutlery with cheeky fish knives in gold with a green eye. The textile designer Ingrid Dessau drew the strict pattern for the linen tablecloths which were made of closely-woven, unbleached cloth. The idea behind this new table setting was to display Swedish quality and design at its best and to present an ideology that pointed out the need for beauty in a world that was too much concerned with figures and profits. Each part of the setting should stand by itself and not be unnecessary decoration.

The dinner would begin with soup in a plate that was placed on a green under-plate. The green would symbolise spring, the continent of America and the science of physics. Africa, summer, red and chemistry were the next combination, Asia, autumn, yellow and medicine the next and Europe, winter, blue and literature the last, for the dessert. There was enormous interest in these beautiful objects. The Nobel Foundation would keep 100 sets, the King and Queen were sent three dozen sets as a present and the remainder would be sold to the public. It was expected, however, that the sets that the Nobel guests ate from and on which the Nobel crest was engraved in various ways would fetch a higher price than the other parts. There was a built-in collector's value in the articles and it was expected that a complete cover set would sell for 5,000 kronor. The following year, it would be only the 88 people at the table of honour who would have the pleasure of eating from this set as the Nobel Foundation would only buy 100 sets. The tablecloth on the table of honour was 210 cm broad and 39 metres long (actually it was 3 × 13 to make it manageable). The napkins were 75 × 75 cm and decorated with the portrait of Alfred Nobel. A very fine thread was needed to weave these napkins with 35 threads per centimetre, which allowed the smallest details in Alfred Nobel's face to be seen. The other tablecloths were only 150 cm broad. The salt cellar in the new set was in the form of a pearl boat. At each cover there was a battery of glasses, five different glasses and a sixth with the coffee.

Two of the glasses had gold stems and one had a gold ball, while the white wine and brandy glasses had a green and a blue stem respectively. Eight pieces of cutlery were needed for the meal and for the dessert they were in shining gold. On December 11, the King and Queen had their own set of the Nobel porcelain with their own monogram — except for the coffee cups. The King thought that would be a bit too much. It was presumed that 3,000 complete sets of the new service had been produced. Each cover cost about 5,800 kronor. The Nobel Foundation ordered 100 sets for the 1991 banquet for the 90 people at the table of honour and borrowed 1,218 sets from the manufacturers. The total number produced was enough for the City Hall to have 1,318 sets, and the stores Svenskt Tenn, NK and Duka Kungsgatan/Sveavägen had a number for sale on December 11. All of them were stamped Nobel Jubilée. After the Nobel banquet, 1,218 sets were sent back to the factories for inspection. The porcelain that had no defects were sent out to various shops for sale and newly produced porcelain to replace the faulty items was stamped Nobel Jubilée. Thus all in, all including the Nobel Foundation's 100 sets, there were 1,318 Jubilée-stamped covers.

Another innovation was that the festival staff was new, made of silver and decorated with a glass City Hall tower. This staff was a gift from the Nobel Foundation to the City Hall in grateful thanks for housing so many successful banquets. The Nobel Banquet gave a fantastic opportunity to stimulate contacts among scientists from different countries. All the laureates who were still alive were invited, and four glass obelisks had also been made to place on the table of honour.

The reason for doing more than usual was not only because it was the 90th anniversary. The Nobel Foundation wanted to show the whole world that it was in Stockholm that the real banquet was held, since Nobel banquets were held in various places in the world on various dates.

There were 1,300 guests in the Blue Hall, including both old and new laureates. Parallel with the banquet in the Blue Hall, the Nobel family organised a banquet for 400 guests in the Hall of Mirrors at the Grand Hotel, the venue of the first banquet in 1901.

The figure four had been the starting point for the banquet: four continents, four colours and four seasons. Each course was introduced by dancers and specially composed music related to each continent. Since there were four courses instead of three, it was important to keep strictly to the timetable. During the two-and-a-half-hour-long dinner, all the guests should eat four courses, and at 9.30 on the dot the coffee should be on the table. Previously, the dinner had lasted four and a half hours, but the shorter time was in part thanks to Prince Bertil, since he had pointed out that it was difficult to sit still for such a long time.

At the signal from the restaurateur, the wine was served and a few minutes later, at 7.05 exactly, the Chairman of the Nobel Foundation proposed the traditional royal toast. The King responded at 7.07 with the toast to the man who was the source of the whole event — Alfred Nobel. Just before 9.30, it was time for the laureates' speeches, by which time the dinner had lasted two hours and thirty-five minutes. A costume carnival preformed by the Stockholm students began at midnight.

The cover price for those not specially invited was 1,500 kronor. The serving was choreographed by Ivo Cramér to the accompaniment of music by Karl-Axel Falk and was based on the figure four, the same number as the original number of prizes awarded in Stockholm. It was all framed by the beautiful flowers that were flown in from San Remo, between 500 and 600 kilos of them. The Stockholm Student Choir sang to the wintry dessert, after which the laureates spoke for five minutes each.

The food was prepared in the kitchen above the Blue Hall and was sent down in two lifts. During the dinner, the Golden Hall acted as the serving room. At 10 o'clock, when the ball was due to begin, it was empty. No one who did not know better could imagine that two small rooms were filled to bursting point with dirty plates. It was only when the feast was over that it was possible to reach the lifts and send up the dirty plates and so on. All in all, 6,500 glasses and 5,200 plates were used, as well as a large amount of cutlery. There were 30 chefs, 280 waiters and waitresses, 20 dish-washers, 6 wine waiters and 12 headwaiters. Most of the staff were brought in from other restaurants all around Sweden since it would have been impossible to arrange an event like the Nobel Banquet with the ordinary staff at the City Hall restaurant, but there was no difficulty in recruiting people since this was a Banquet with a capital B in the restaurant world. Those who were selected to work at the Nobel Banquet were awarded a diploma for their contribution. During the afternoon of the banquet, the dessert wine was uncorked and at about the same time as the last bottle was opened (about a thousand were needed) the serving staff began to lay the tables with the new tablecloths and the new service.

The new Nobel service came up to expectations at the banquet. Since it was composed around the theme of the figure four and the menu had been changed to four courses instead of three, the banquet was as much a feast for the eyes as for the taste buds. The knives and forks were described as easy to handle, the food was even more delicious than ever before and the wines were perfect. The soup was brought in to the accompaniment of trumpets and drums played by six heralds carrying high standards in the form of ships loaded with amaryllis. The blue ice-cream, this year illuminated by flickering Advent candles, was accompanied by equally touching Swedish songs performed by student singers. The traditional fireworks sparkled down the City Hall walls in the form of four Bengal lights. Peter Frohm, who had worked at the City Hall before, was now working as head chef at the Sara Hotel in Bejing, China and he had the Nobel menu telephoned to him. That meant that the 90th-anniversary menu could be served at this hotel only 17 hours after it was served in Stockholm.

This jubilee Nobel Banquet would undoubtedly go down in history as a magnificent event that none of the guests would forget in a hurry. The festival in the Blue Hall was a magnificent show of light in the winter darkness. It concluded with a grandiose firework display, a gift from the city of Valencia in Spain that weighed 1½ tons.

The City Hall restaurateur revealed that it was not only on Nobel Day that the Nobel dinner was served at the restaurant. It was normal for the City Hall restaurant to get orders during the whole of the rest of the year for this splendid meal. They did not need to be for that year's dinner, they could just as well be for previous Nobel dinners.

In 1992, the 1991 Nobel Service was awarded the Swedish Design Prize. This service had now been enlarged with a salad bowl and a teacup. The new Director of the Nobel Foundation, Michael Sohlman, decided that over a five-year period the Foundation would buy the Nobel service for all the 1,300 guests at the banquet.

Care, attention and willingness, together with competence, were the most important factors when planning the banquet. These important ingredients were in the hands of 20 chefs, 15 dish-washers, 200 waiters and waitresses, 8 headwaiters and 5 wine waiters, and of course the restaurateur. The serving staff included 21 catering school students from Alingsås.

The tables in the Blue Hall were decorated with white Christmas roses, and green garlands combined with geometric flower arrangements in white

Ingrid Carlsson (the wife of the then Prime Minister of Sweden) is served Glace Nobel at the table of honour in 1986. It now has a new design; instead of the high, square tower, it is now round and looks like a two-tier cake. © The Nobel Foundation/Boo Jonsson

The Glace Nobel parade in 1984 led by a folk-music fiddler. A great deal of professional skill is required of the serving staff when they carry the heavy dishes of ice-cream down

the stairway in the Blue Hall, also bearing in mind the breadth of each step. © The Nobel Foundation/Boo Jonsson

Glace Nobel served at one of the tables in the Blue Hall in 1985. The picture is an example of how the logistics at the banquet function, and a table at which both adults and children have been placed. If you look closely, you can see that the adults still have some wine left in their glasses and the children a soft drink. Evidently the age of the guests is of no importance when it comes to getting the caramelised N from the ice-cream since everyone in the picture has managed to get one each on their dessert plates. I wonder how that could be since every Glace Nobel has one N and each dish is supposed to be enough for ten persons. © The Nobel Foundation/Boo Jonsson

The custom of test-eating various proposed menus for the Nobel Banquet was also followed before the 90th anniversary of the prize-giving. Here we see the winning proposal for the first course, nettle soup with quail's egg served on the specially designed Nobel service. © IMS bildbyrå

The ice-cream parade at the 90th anniversary banquet in 1991, flanked by neon lights. The Glace Nobel has again been given a new design consisting of a large tier of ice-cream

and parfait and a small tier on which a large candle has been placed. The candy floss and petits fours have been retained. © The Nobel Foundation/Boo Jonsson

Serving the dessert in 1991 at one of the tables with a student choir in the background. The waitress is demonstrating considerable professional skill considering the small space between each guest. She has also managed to serve all the portions and still have some ice-cream left in the middle of the dish — and the candle is still burning. © The Nobel Foundation/Boo Jonsson

For the 90th anniversary, a variant with individual portions of the dessert was developed. It looked like the picture here, served on the beautiful dessert bowl in the newly designed Nobel service. © Photographer: Holger Edström

The 90th anniversary of the Nobel Banquet attracted great attention. Here we see the table laid with a coffee cup and a glass of brandy for the media. However, brandy was never served at the banquet. © Photographer: Holger Edström

Right: The result of the morning's work in 1992. A detail picture of one of the covers in the Blue Hall when everything was laid out. © The Nobel Foundation/Boo Jonsson

20
A

A detail picture of the table of honour in 1996. The dessert has just been served and this year the caramelised N has been reintroduced. The guest on the far left has one of them on her plate. There are no longer any petits fours on the same dish as the Glace Nobel;

instead, they are served on silver dishes of their own which are placed on the table.
© The Nobel Foundation/Boo Jonsson

Serving the dessert in 1992. It has the same design as at the 90th anniversary the previous year. The caramelised N which used to decorate the Glace Nobel has been replaced by an N made of almond, and now there is an N for each guest, to the children's great delight. © The Nobel Foundation/Boo Jonsson and yellow adorned the Hall itself. The serving processions were preceded by dancers from the University College of Dance, with its professor in the lead. Apart from the serving staff, there were also the Royal Dramatic Theatre's young medieval singers in the processions. The climax of the entertainment was the Pyramid Dancer's solo in a winged dress which she performed immediately before the Glace Nobel was served. This year's festivities were designed with the help of university staff in the fields of handicraft and dancing.

The dinner lasted about three hours. The toastmaster fainted twice and right in the middle of the Chemistry laureate's speech she lay flat on her back on the stairs in the Blue Hall before she was helped to her feet. An unfortunate waitress dropped a tray of the Nobel service on the Blue Hall stairs. The banquet was attended by 1,268 guests. The Nobel ice-cream was considered the most stylish of the dishes, being served from a dish with an obelisk-like candle in the centre, but some of the serving staff thought that it was impossible to handle. The menu was also described as being more continental than before since no potatoes were served with the main course, but there were vegetables fried in butter instead.

While the guests gathered, a Baroque Ensemble of students from the College of Music played music by Handel and others. When the standard bearers marched in, they played the first movement from The Drottningholm Music by J H Roman. Before the first course was served, a procession of dancers from the University College of Dance entered the scene together with a group of jesters. The white-clad dancers wore garlands of pine branches which they used to decorate the stairway between the Blue Hall and the Golden Hall. The main course, too, was introduced by jesters and dancers who carried large boards covered in flowers, yellow and white gerberas, carnations and yellow and white lilies. The Glace Nobel was served with candy floss and the letter N in almond pastry. The final entertainment at the banquet was performed by the Stockholm Student Choir. After the dessert, the laureates gave their speeches. The Literature prize-winner did not eat the ice-cream and chose instead to have fruit for dessert; nor did he drink any wine because of his diabetes, only non-alcoholic drinks throughout the banquet. Some catering school students from Lidköping were among the serving staff, the youngest of them only 16 years old.

In 1993, too, a special brains trust of art historians were asked to plan the contents of the banquet. The theme was a Scandinavian winter saga.

The Nobel service from the 90th anniversary banquet was in white, gold, blue and yellow while the decorations were orange and green. In tune with the evening's theme, the table of honour was decorated with hazelnuts, club moss and 15 kinds of Swedish winter apples of all conceivable colours. The other tables were decorated with Christmas roses. Before the dessert, Barbara Hendricks and the choir Orphei Drängar gave a much acclaimed concert. The choir were dressed up as waiters and they burst into song among the tables, praising the wine. The Glace Nobel was luminous and was served with candy floss on illuminated blocks of ice accompanied by neon rods as cold as icicles. After the dessert, it was time for the laureates' speeches, beginning with the Literature prize-winner. This was the end of the formal part of the banquet and the guests left the table 20 minutes after the appointed time. For the first time, the guests were allowed to mingle in the Prince's Gallery, where the Haga Chamber Trio entertained them by candle-light. While the royal party withdrew to converse with the laureates, the ball began in the Golden Hall. The food that was left over from the dinner was served as today's special at the City Hall restaurant the day after the banquet.

There were 1,278 guests and the atmosphere at the banquet was ceremonial but relaxed. The fact that the Nobel Banquet was by tradition well programmed did not at all mean that it was not a world-class feast. The reason was that the whole arrangement had been turned into an art form of its own based on both rhythm and feeling.

This year was the 10th anniversary of Glassbruket's first delivery of the Glace Nobel. According to this supplier, a Glace Nobel should be narrow and high yet natural. The raw materials were also important: no preservatives, artificial flavouring or false cream. Four to five days before Nobel Day, Glassbruket began to spin the sugar for the 140 Glace Nobel. Three days before, the ice-cream itself was made, after which it was taken to the City hall in a refrigerated lorry. When the main course was carried in, the staff began to put out the Glace Nobel. They had about 35 minutes before it was time to take the ice-cream in. When the finishing touches had been completed, each Glace Nobel had been given two minutes' attention before it was carried in. The choice of wines depended on the tenders that all the wine merchants who were interested had sent in. The Nobel Foundation in cooperation with the City Hall restaurant then selected what they preferred. The main criterion was that there had to be a good balance between the food and the wine. Six members of the Nobel

Foundation test-ate the menus that were proposed on November 3rd, after which the menu was finalised. This year, the City Hall restaurant bought the Nobel service, which is used when the guests there order the Nobel Banquet menu.

Now it was 1994 and the Nobel Banquet was seen as a meeting place for all kinds of people. The banquet reflected the meeting. The fact that 1,300 people met and ate together was a fascinating thought. The Nobel Banquet was also a feast in the name of science, all in accordance with the spirit of Alfred Nobel. It should never be allowed to become a publicity stunt. The 1,300 guests were selected according to specific rules; they should have a connection with one of the laureates, be of importance to the Nobel Foundation or belong to the academic community. The day before the banquet, the door to the attic where the Nobel service was kept was unlocked by the restaurateur. The serving staff then put on white cotton gloves, took out their measuring sticks and began to lay the tables. Since the Blue Hall is not symmetrical, each table has to be placed individually; moreover, the distance between each cover must be the same and each glass, knife, fork and spoon has to be polished. The students' interest in taking part in the banquet was enormous. Taking part in the students' ticket lottery cost 50 kronor per couple and the invitations were sent out to universities and university colleges in October. Between 1,500 and 2,000 applications were sent in. When all the envelopes had been opened and each application had been given a number, the 30 members of the club-masters committee threw dice for a whole day to pick out the 200 lucky students who won tickets to the banquet. The names of the winners remained a deep secret since the tickets were so sought after that those who had won them ran the risk of being almost harassed by hopeful buyers. Even though the tickets were personal and could not be transferred or sold, the black-market price was said to be 8,000 kronor per ticket.

When the Nobel guests arrived at the Stockholm City Hall, they were welcomed by Örjanslåten, a chime from the carillon in the City Hall tower. Flaming torches were placed round the City Hall and fires lit up the edge of the quayside. Sea scouts from Nacka stood with flaming torches in the courtyard.

The table setting was described as beautiful and stylish; the idea of decorating the tables with red and yellow Swedish apples instead of flowers was very popular, and was even described as magnificent. Then there were the candelabra with 244 candles. The music at the banquet was specially composed by Sven-David Sandström, with Eric Ericson as the conductor. It was felt by some people to be courtly and elegant while others thought it was cold and rather weird.

There were 1,278 guests plus 250 students. One "guest" boasted that he had managed to get in without a ticket or any cost to himself for 17 years. The cover price was 1,600 kronor. The banquet began with a psychedelic light show in which stars and undefinable celestial phenomena flew across the ceiling high above the heads of the guests in the Blue Hall.

At 6.25, the guests were requested to take their places for the entry march, Marche Turque by Jean Baptiste Lully. At 6.45, there were introductions in the Prince's Gallery in the presence of the royal family and the ambassadors of the laureates' countries; at 7.07 the wine was served at the main table, at 7.10 the royal toast was proposed by Professor Bengt Samuelsson, Chairman of the Board of the Nobel Foundation and two minutes later His Majesty the King raised his glass in honour of Alfred Nobel. At 7.15, the first procession with the first course descended the stairs accompanied by the Stockholm Chamber Brass; at 8 o'clock, the second procession carried in the main course to the accompaniment of Eric Ericson's Chamber Choir and the Stockholm Chamber Brass; at 8.45, Swedish folk calls and dances performed by Lena Willemark and Peter Jonsson before the dessert; 9.25, Coffee and tributes to the laureates by the Swedish students carrying their massed standards. A fanfare was sounded before each laureate's speech. The speeches were held in the following order: Literature, Physiology or Medicine, Chemistry, Physics, Economic Sciences; 9.50, Eric Ericson's Chamber Choir entertained on the stairway. At 10 o'clock, the guests left the table to the tones of Concerto by Benedetto Marcello. The ball was opened by this year's dance band. The royal family received the laureates.

This year's dinner required 330 bottles of red wine, 320 bottles of champagne, 450 metres of linen tablecloths, 2,576 silver knives and forks, 5,152 glasses and plates, 140 Glace Nobel and 253 serving staff.

The food was described as excellent, light and elegant. The ice-cream was carried in preceded by blue neon rods more than one metre high. The ice-cream was decorated with candy floss and an N. The banquet was felt to be outstanding and a total experience. An innovation was the N-shaped Croissant Nobel instead of the usual Nobel roll.

The entertainment also included a girl singer who sang folk calls accompanied by a dance by male dancers. A folk call (Sw. kulning) is the mountain call of girl cowherds sung at a very high pitch. Eric Ericson's Chamber Choir and the Stockholm Chamber Brass sang and played during the dinner and the Philharmonic Salon Orchestra played dance music in the Golden Hall.

Each course was brought in with a procession led by Gunnar Cyrén's heavy silver and crystal glass staff decorated with golden boats filled with gerberas and spruce twigs. There were also moss-covered squares and the shining blue neon rods from Öland. However, there was a 20-minute delay when the Glace Nobel was to be brought in. The cause was unclear but there was a report that the ice-cream had begun to slip off the dishes before they were carried in.

1995 was the year when Crown Princess Victoria made her debut at the Nobel Banquet. There were 1,302 guests and the evening's entertainment centred on the 18th-century Swedish poet and composer Carl Michael Bellman, who died in 1795. The stewards now had to eat in another room — but it was not any old room but the Prince's Gallery with its beautiful view across Riddarfjärden. This year no one on the reserve list had been invited because of the large number of guests.

The more than 1,300 guests were welcomed to the City Hall by the tones of Örjanslåten and a large brazier burnt on the parterre by the quayside. The entry march into the Blue Hall was introduced by a Bellman-arranged fanfare from two trumpets. The long tables had been laid with porcelain decorated in green, yellow and blue. Baskets of Swedish-grown pears decorated the tables. As was the tradition, the three courses were preceded by processions, in the case of the first course a modern, choreographed dance with four women and one man dancer from the University College of Dance in Stockholm. When it was time for the main course, Bellman's music took over. Students from Adolf Fredrik School's music classes sang and performed Bellman compositions at the same time as they decorated the Hall with 7,000 carnations of all colours. The third procession before the dessert was a musical illustration of the street crowds in Bellman's time. Four soloists from the University College of Opera together with the Chamber Choir and Adolf Fredrik's Boys' Choir performed. The food was as usual described as superb. The ice-cream was served on blocks of ice. The Glace Nobel, which was coloured yellow, orange and red, was called sunset and it matched the flower decorations of carnations, lilies and gladioli. It was crowned with a sugar N. The speeches of thanks by the laureates followed and the banquet came to an end at about 10 o'clock. Then came the ball in the Golden Hall in a more exhilarated mood which went on until 12.30 to the music of two orchestras from the University College of Music.

At the beginning of the banquet, the guests glanced around them and leafed through their seating list in a rather preoccupied manner. An elegant murmur rose to the ceiling and after the toasts, when the champagne had got into their bloodstreams, the atmosphere became more relaxed. By the end of the dinner, people were talking to each other with lively body language across and round the tables and were really having fun. The food was considered so tasty that people would have licked their plates if they had dared to. The Literature laureate said that it was a bigger and better event than he had expected and that it functioned as an initiation rite so that you were not the same person when you left as you were when you came in.

There were 3,000 eager applicants for the coveted work as serving staff at the Nobel Banquet; 250 of them were selected and one of them was a catering school student in his last year who had chosen the Nobel banquet as his special assignment. Getting into the staff was more difficult than he had reckoned on, however, since the restaurateur preferred to take people who had finished school and worked for a few years before joining the Nobel Banquet team. But after a letter of recommendation from his teacher, this lucky student was allowed to come and serve. The waiters and waitresses walked at least three kilometres during the banquet. There was also a student chef among the otherwise professional kitchen staff. The vegetarian food was kept hot in a special oven in the Nobel kitchen and the dietician had prepared a special menu for the guests who had requested this alternative.

The year 1996 was the centenary of Alfred Nobel's death and the banquet was arranged to celebrate this anniversary. A ticket to the banquet cost 1,650 kronor and there were almost 1,300 guests, 200 of whom were students; 83 of the guests sat at the table of honour.

The guests at the banquet were to be given the impression that they were sitting in a garden. This was achieved by placing a jungle of palm trees behind the grand stairway in the Hall. There were also yellow roses, butterfly orchids, flamingo flowers and decorative rust-red cabbages in golden, boat-shaped bowls along the east wall. It was all meant to be very luxuriant and beautifully scented. Above the colonnade along the east wall, a feeling of hanging gardens was created by using long, French tulips, roses, palm fruits, ferns, chestnut vines, proteas and rose-pink orchids. The tables were decorated with parrot tulips, small roses and tropical fruits like rambutan, kumquat, mangosteen, feijoa, physalis, baby pineapples, grapes and gilded nuts. Finally, the stairway was decorated with champagne roses and foliage from dark-green pot plants.

The Nobel Banquet menu was composed by six Chefs of the Year (chefs who in various years had won the Gastronomic Academy's dairy medal). These six chefs, together with the City hall restaurant's own staff led by the head chef, were the ones who prepared the food for the Nobel guests. During the summer, the Chefs of the Year had been asked to send in suggestions for Nobel menus and during the autumn they had cooked their menus, which the Nobel Foundation, together with the head chef at the City Hall restaurant, had test-eaten, discussed and finally decided on for this year's menu. The Chefs of the Year who took part thought it was great fun and a great honour to be given this assignment. The reason for the original request lay in the great success the Chefs of the Year had both in Sweden and abroad. The City Hall restaurant also wanted to make use of the skills and inspiration that these chefs had succeeded in communicating. This was also a great opportunity to market contemporary Swedish gastronomy.

The first criterion was that the food should be good; other criteria were to present the best of Swedish food — ingredients, cooking methods and traditions. Classic, old-style techniques would be blended with modern technology. This year's Nobel bread was made with the four Swedish grains, wheat, rye, oats and barley, which were strewn over the rolls. They were so delicious that they did not need any butter. Seven different kinds of cakes were served with the Glace Nobel, a tribute to the traditional Swedish afternoon coffee party, though in modern form. The cakes were tiny so that the guests would manage to taste all of them.

The timetable for the banquet looked like this: 7.00 Guests at the main table walk in procession out onto the balcony to the tones of Marche Triomphale by Jacques-Nicolas Lemmens; 7.09 Royal toast proposed by the host, Professor Bengt Samuelsson; 7.10 Toast proposed by the King in memory of Alfred Nobel; 7.14 Hors d'oeuvre served accompanied by the Polonaise from Eugène Onegin by Peter Tchaikovsky; 7.50 Twelve torch bearers lead the serving staff with the main course accompanied by Persian

March by Johann Strauss Jr; 8.50 Musical soiree followed by the dessert; 9.45 Coffee followed by the laureates' speeches; 10.10 March out, followed by the ball in the Golden Hall. The Blue Hall was filled with cultivated conversation during the dinner and there was a friendly atmosphere.

The more than 200 waiters and waitresses could not take more than four minutes to go from the Golden Hall to the Blue Hall carrying the plates. The food was prepared, sent down in the lift and placed in ovens exactly 13 minutes before it should be served. The wine waiters took the empty plates out with them. The serving staff should keep their exact positions throughout the dinner and they walked about five kilometres during the evening. Five people worked exclusively on uncorking the bottles. In addition to the restaurateur, the head chef and the room manager and his assistants, 8 headwaiters, 210 waiters and waitresses, 20 chefs and 20 dish-washers and carriers were employed.

There were 1,346 guests at the 1997 Banquet, which meant that the City Hall Library had to be opened for some of them. The flower arrangements in the Blue Hall consisted of white orchids and white carnations, a gift from San Remo. The carnations decorated boxes in which ice sculptures had been placed. The theme was winter and all the flowers were white: carnations, orchids, kochia, calla lilies and amaryllises. It was students in their final year at the Florist High School in Norrköping who arranged the flower decorations: 200 compactly grouped carnations decorated the table of honour. Magnificent, high, gilt-bronze dishes held baby pineapples, wicker orbs and balls drenched in sugar. Along the walls hung white orbs like snowballs with spruce branches, flamingo flowers and fragments of broken glass that glittered silver in the lights. The price of a cover was 1,650 kronor per person. The guests were welcomed by 500 blazing torches outside the City Hall. The music played at the banquet was mainly from the 20th-century and specially composed music by Georg Riedel.

The seating plan at the table of honour followed a certain pattern. In the middle of the table sat the King and the Queen, and the rest of the places were arranged accordingly. The King's dinner partner was always the woman Nobel prize-winner, and those years when there was no woman prize-winner, the wife of the oldest prize-winner took that place. The Queen's place was opposite the King's and she always had the Nobel Foundation host as her dinner partner. It was also he who walked with the Queen in the procession down the stairway to her place. The places closest to the royal couple were distributed among the laureates, and there was a so-called ranking list for these places. The laureates sat in a certain order in relation to the royal couple; closest was the Physics prize-winner, followed by Chemistry, Medicine and Literature and furthest away the Economics prize-winner. This seating followed the order in which Nobel had listed the prizes in his will. Laureates from previous years and their partners always had a place guaranteed at the table of honour. Then there were a number of government ministers. The table of honour had 90 covers. It was also decorated with six ice sculptures as well as with bowls of nuts, chestnuts and sweet-smelling oranges spiked with cloves.

The timetable for the banquet looked like this: 6.35 The guests were requested to take their places. 6.45 The laureates and other guests of honour were presented to the Swedish King and Queen in the Prince's Gallery. 7.00 The guests at the main table walked in procession to their places to the accompaniment of organ and trumpets music. 7.10 The King proposed a toast in memory of Alfred Nobel and a minute later a number of press photographers were allowed in to the main table to take photos for four minutes. 7.15 The hors d'oeuvre was served to the accompaniment of music played by Södra Latin High School's Brass Ensemble. 7.56 The main course was served with more brass-band music. 8.25 Musical entertainment for 20 minutes while the dessert was dished up. 9.45 Coffee and the tribute to the laureates from the Swedish students with their standards, followed by trumpet fanfares and the laureates' speeches. 22.10 The guests broke up for the ball in the Golden Hall to the music of Lars Kvensler's orchestra. The King and Queen received the laureates in the Prince's Gallery.

The serving staff consisted of 240 people, four of whom were students taking the hotel and catering programme at Vipenholm School. In addition, there were 250 people providing the entertainment, which meant that there had to be a co-ordinator who timed every event so that nothing went wrong. The entry march should end exactly when the first course was brought in, no earlier and no later, so that the musicians and the waiters and waitresses should not collide in the exits and entrances. The most difficult job for the serving staff was the ice-cream as it had to have the right consistency, not too hard and not too melted. It was also tough to have to stand and wait for the signal to march in. It was always a waitress who served the King and a waiter who served the Queen. Incidentally, the King always used to serve himself and not let the waitress serve him. To be a wine waiter, you had to be able to carry four bottles without your hands aching; these waiters served the champagne first, then the wine, the dessert wine and mineral water in the fourth glass.

The previous year's concept of a team of Chefs of the Year who put forward suggestions for the menu and then prepared the food together with the permanent staff in the Nobel kitchen was followed this year too. When the guests marched into the Blue Hall, the staff had been working on the food for three days. Ten months earlier, discussions about what food should be served had got going and the first menu was test-eaten in September. The silver dishes from which the ice-cream was served were garnished with luminous light rods and white candy floss. Four hundred litres of coffee had been brewed to go with the seven kinds of cakes. The banquet and the food were described as beautiful.

The leaders of the Swedish political parties had always been invited to the banquet ever since 1992, but now in 1998 a new system was introduced whereby they would be invited every other year. The reason for this alternating system was that the pressure from foreign guests had increased while the number of seats in the Blue Hall was limited. The cover price was 1,700 kronor per person. It had been decided that the laureates could bring 12 relatives to the banquet over and above their closest family. Two hundred of the 1,286 guests were students. The reason why the number of guests was lower than the previous year was that some of the places were not so good to sit at.

This year's flower arrangements were based on roses and the Blue Hall was decorated with roses in four colours which had been arranged by florists from Norrköping High School directed by Jim Grundström. Love was the evening's theme; in bowls on the table of honour lay what seemed to be giant roses but which actually consisted of several hundred rose petals. Here and there stood golden balls of wild-rose thorns glued together as a reminder that even love has its thorns.

At 6.30, the guests were requested to take their places at the tables. The laureates and the other guests of honour were received by the King and Queen in the Prince's Gallery. At 7 o'clock, the guests seated at the table of honour walked in procession down the stairs in the Blue Hall and took their places The royal toast was proposed by the host and the King responded by proposing a toast in memory of Alfred Nobel. The banquet ended at about 10 o'clock and then it was time for the ball in the Golden Hall. The entry of the Glace Nobel was a hit when it was brought in in procession with the lights dimmed in the Hall and the dessert dishes lit up with red lights. This year the ice-cream was blackberry parfait embedded in white chocolate and crowned as usual with the letter N. The dinner lasted three hours and twenty-five minutes but the time seemed to fly as feelings of pleasure and well-being quickly spread among the guests.

Again this year, the concept of the Chef of the Year in cooperation with the permanent staff was followed when composing and preparing the menu. Twenty chefs worked on the food and 210 waiters and waitresses took care of the serving. As for the menu, there were 15–20 guests who had requested a vegetarian alternative, which of course was provided. What was considered "fine food" was controversial and the restaurateur related that there had been certain comments in the year they chose to serve marinated cod — although he hastened to add that it was not deep-frozen blocks of cod. The reason for choosing cod was to present Swedish gastronomy.

One criterion or minimum requirement to become a member of the serving team at the banquet was to have had ten years' experience of serving. If you had that experience, you were selected to serve at a banquet like the Nobel Banquet in terms of the number of courses and guests. After that, you were told whether you had been selected to serve at the Nobel Banquet. When the banquet was over, the serving staff had done 14 rounds of 192 metres, that is, walked three kilometres carrying heavy serving dishes and plates. To judge by the newspapers, it seemed as though there was no longer any discussion about the status of the Nobel Banquet. It is **the** great feast of the year and nothing finer exists.

In 1999, the feeling was that the Nobel Banquet should be a feast for the laureates and an event that should have a more academic character than before. It should act as a meeting place for foreign and Swedish research. No representatives of Swedish business life were on the invitation list except those who were sponsors and donators in some scientific field. About 40 per cent of the guests were the same year after year. Standing invitations went out to the 361 members of the Royal Academy of Sciences, the 18 members of the Swedish Academy and 50 people from the Karolinska Institute. The government, the Speaker of Riksdagen and the royal court also had standing invitations. Seating all the guests at the right tables took a whole weekend.

1,270 guests attended the banquet and now the leaders of the political parties were invited; they had not been invited the previous year. This the last Nobel Banquet of the last century led to a demand for tickets that was greater than ever before. A ticket cost 850 kronor if you were backed by a company and 1,700 kronor for a private person. The laureates could invite 15 guests each to the banquet itself.

The Nobel Committee had started to cooperate with the Chef of the Year Society. All the chefs who have been elected Chef of the Year are members of this society. Since 1996, some of these chefs have been asked if they would like to help create Nobel menus for the coming banquet. The Chefs of the Year who were asked accepted the invitation, though the reward for their work would be quite simply the honour and a Nobel diploma, of course. They felt honoured to have been asked and to have been allowed to take part. What was to be served had been planned since the spring, and in September there were four menus ready to be test-eaten, only one of which would be chosen. The food had to be elegant, tasty and Swedish; 30 chefs took part and it took then three days to prepare the Nobel food. The menu was protected as if it were a state secret; nothing about what was to be served was to leak out.

The theme for the evening was Optimism and Faith in the Future, a Bridge to the Future. The main flower in the arrangements was Gloriosa Superba, the red climbing lily, backed by mimosa, lemon-yellow calla lilies, hypericum berries and steelgrass. There was also the decorative climbing plant Asparagus Asparagoides in the Prince's Gallery Students from St Göran's High School made the flower arrangements.

At 6.45, the laureates and the other guests of honour were presented to the King in the Prince's Gallery and a quarter of an hour later the guests at the table of honour walked out in procession on the balcony through Snäckan to enter the Blue Hall, accompanied by trumpets and organ music. Nine minutes later came the royal toast and one minute after that, the King responded with the toast in memory of Alfred Nobel. At 9.11 there were four minutes of photographing at the main table, two minutes for the left-hand side and two minutes for the right-hand side. Then at 7.15, the first course was served to the accompaniment of music played by The Orchestra conducted by Olov Olofsson. At 8.10, the main course was served to music by the same orchestra and at 9 o'clock, it was time for 20 minutes' musical entertainment by the mezzo-soprano

Malena Ernman, the actor Magnus Krepper, Sofia Lindqvist and The Orchestra. Then it was time for the dessert. At 9.55, coffee was served and the Swedish students paid their tribute with their standards before the laureates gave their speeches of thanks, each of which was preceded by fanfares of trumpets. Finally, the guests left the table at 10.20. The ball in the Golden Hall took over with music played by The Orchestra while the royal couple received the laureates in the Prince's Gallery.

The atmosphere at the banquet was characterised by Tradition and Ritual. The first course was already on the table, so all the waiters and waitresses had to do was lift the covers from the beautiful Nobel dinner service. The food tasted heavenly. This was noticeable in the Hall because the guests' conversation subsided when they started to eat and sighs of satisfaction were heard from time to time during the dinner. A compote of passion fruit and crushed pineapple was served with the dessert. The rest of the food was described as very Swedish, and the menu was thought by many of the chefs to be the best so far. 210 waiters and waitresses carried in the food, the delicacies, 140 of them served the food and 55 the wine. During the hours that the banquet took, almost a whole year of planning disappeared down the guests' throats and all that was left were the dirty dishes which it took 19 people four days to wash up. The banquet required 5,276 glasses, 4,000 plates, 7,900 pieces of cutlery, 280 kilos of meat, 150 kilos of potatoes, 700 litres of sauce, 300 bottles of champagne, the same number of bottles of red wine, 1,500 litres of mineral water and 270 litres of coffee.

In 2000, all the three children in the Swedish royal family attended the Nobel Banquet for the first time. They were placed at the table of honour. There were 1,280 guests, 200 of whom were students. Each laureate could bring 13 guests to the banquet and most of them filled their quota. It took a whole weekend to place all the guests since it was the aim of the Nobel Foundation that they should enjoy both a scientific and a cultural exchange from the banquet. From the government came the Prime Minister, the Foreign Minister, the Minister of Education and the Minister of Cultural Affairs; the leaders of the opposition were not invited on account of the alternating schedule. The tickets cost 1,735 kronor each. Outside the City Hall, 500 torches burned to welcome the guests.

Now the Nobel Banquet was described as the feast of all feasts, the most sought after of all banquets. The preparations in the City Hall were described as a real military campaign led by the restaurateur, who directed the 210-man strong serving staff with discreet signals. Pedometers, stopwatches and measuring sticks all played their parts in the preparations. The dinner service that was made for the 1991 banquet had become almost too popular by now; a certain wastage of plates and so on was noticed. For that reason, the serving staff had been asked to lay out the cutlery as fast as possible. This year, there were eight catering students from Allé School's hotel and restaurant programme among the staff. The school had applied four times before but without success. These eight students were honoured to have been chosen as this was one of the most momentous things that a waiter/waitress could experience in Sweden.

Seven Chefs of the Year joined up with 15 local experts to prepare the Nobel food. The chefs who were the seven Chefs of the Year pointed out that the permanent staff knew the routines and those who came in as guest chefs had new ideas, which were appreciated as everyone involved was able to exchange experiences and develop their skills.

The theme of the Banquet was Time. The Blue Hall was decorated as a garden and among the arrangements were the 5,500 carnations and 3,500 sprigs of mimosa donated by San Remo, but also Swedish lilies of the valley and moss. The dominant colours were green, white and pink with dashes of red and blue. The Prince's Gallery was decorated with long bridges of artistically arranged flowers and the table of honour with small, delicate gardens made of moss and small lilies of the valley, anemones, thistles, roses and Christmas roses. In the large, high and impressive golden bowls there were pomegranates, rose hips, figs, persimmons, chestnuts, walnuts and lemons. The Blue Hall had also been decorated with illuminated cloud effects in the ceiling and flower-decked pillars looking like trees. The entertainment, dancing and singing, was provided by the soprano Jeanette Köhn, the mezzo-soprano Katija Dragojevic and the playful clarinettist Martin Fröst. A well-behaved murmur of conversation rose from the tables. When the food had been served, it was described as being so beautiful that it was a pity to eat it up. The banquet itself was "The Great, Wonderful, Fantastic Banquet". After a finale, Monteverdi's The Coronation of Poppea, the Blue Hall exploded with great flowers of light and the Glace Nobel was carried in. It consisted of vanilla ice-cream with a sorbet of bitter-sweet lingonberries and spartan country biscuits — very Swedish and delightfully unsnobbish. After dinner, the Ball could begin in the Golden Hall.

The asparagus for the main course in 2001 on ice in the kitchen. © Photographer: Torbjörn Boström

The famous Glace Nobel, Model 2000. © The Nobel Foundation/Hans Petterson

The first course is put out in the kitchen at Stockholm City Hall before the banquet in 2001, which was the centenary of the Nobel Banquets. © Photographer: Torbjörn Boström

The main course is put on its plates in the centenary year, 2001. © Photographer: Torbjörn Boström

The Glace Nobel had a new design for the centenary year, 2001, and was now served as portions. © The Nobel Foundation/Hans Pettersson

Serving the main course in the centenary year, 2001. The picture makes it clear how little space there was between the guests, since they both lean to one side to help the waiter. We can also see an example of the serving staff's professional skills at the banquet. © The Nobel Foundation/ Boo Jonsson

The first banquet after the Second World War was held in 1945. It started with a creamy mushroom sauce. The porcelain is from Bungalow Porslin AB, Stockholm. © Photographer: Torbjörn Boström

The mushroom soup was followed by saddle of reindeer with haricots verts cooked in butter, cream sauce and lingonberries. © Photographer: Torbjörn Boström

The 1945 menu ended with a bombe glacée with no trimmings. Like the menu served in 1920 after the First World War, the 1945 menu set the pattern for the number of courses that would be served in the subsequent Nobel Banquets with the exception of 1991, that is, three. The porcelain is from Bungalow Porslin AB, Stockholm. © Photographer: Torbjörn Boström

The author of this book, Ulrica Söderlind (left) and Lena Eriksson putting out the candy floss for the Glace Nobel in 1990. This was done in the Golden Hall while the guests were seated at the tables in the Blue Hall below. Each dish is decorated with an edge of candy floss, and the Glace Nobel, which is enough for ten persons, is placed in the middle.

The petits fours are laid out on a silver dish of their own in the kitchen before being sent down on trolleys by lift to the Golden Hall below. Each dish contains ten petits fours, which is enough for one Glace Nobel.

When the guests have been served their dessert and the dirty plates have come back, the work of clearing up begins so that the Golden Hall is clean when the ball begins. After

everyone has given a helping hand, they sit down in the kitchen and eat the Nobel menu. This photograph was taken in 1990 and it was very late in the day.

Everyone who has worked at the Nobel Banquet and in the days beforehand is awarded a diploma after the festivities are over, regardless of their job. © Ulrica Söderlind

For the serving staff the timetable looked like this: 3.30, everyone should be in place to receive instructions; 4.00, run-through of the programme by the banquet manager in the Golden Hall; 5.00, rehearsal. At 7.15, the hors d'oeuvre was served. at 7.40, the plates were cleared away, ten minutes later the main-course plates were put on the tables and at 8 o'clock the main course was served; 35 minutes later, the main-course plates were cleared away and at 8.25, the dessert plates were put in place. At 9.10, it was time to serve the dessert and half an hour later those plates were cleared away. At 9.50, coffee was served and at 10.03, most of the serving staff relaxed with a night sandwich in the Restaurant Eken, while about 30 waiters and waitresses stayed on duty. At 10.20, the guests left the table and the tables were cleared. A few of the staff stayed on to serve at the refreshments bar in the Golden Hall during the ball. For the 50 wine waiters, the timetable was a little different. At 6.45, champagne was served to the guests who had taken their places at the tables and at 7.07, champagne was served at the table of honour. At 7.25, it was time for refills. Twenty minutes later, the red wine was served and the glasses were refilled at 8.15. At 8.40, the dessert wine was served with a refill at 9.20. Finally, at 9.53, brandy and liqueurs were served. The banquet required 9,100 pieces of cutlery, 7,800 glasses and 6,500 plates.

The year 2001 was the centenary of the Nobel prizes. Consequently, all the laureates who were still alive were invited. There was no specific terrorist threat against the Nobel Banquet, but the security had been tightened up in any case. About 200 living laureates were expected, among them many Americans. Both the Stockholm police and the Security Police had more men on duty than usual. The reasons were the 9/11 terrorist attack, the war in Afghanistan and the fact that more than 100 former American prize-winners would be attending the banquet. Despite the situation in the world, only one cancellation had been received on this account.

Preparations for this Nobel Banquet had been going on for two and a half years. The total number of guests was just over 1,400, plus 200 students who had dinner in the City Hall restaurant. Many requests to be invited to the banquet had been received, but this year it was impossible to grant them — as indeed it had been in previous years. Those who were invited were, apart from the laureates themselves, their relatives, the 200 students, the royal family and the representative of the government. If you had had an academic career or held an important position in society, you might be considered, too. The Nobel Foundation pointed out that it was important to remember that the banquet was both an academic event and a private party. The seating plan was organised by having the whole list of guests on a computer together with information about their interests and languages, so that the guests could be paired off as well as possible. It took a whole weekend to solve the problem. Who would be sitting next to whom at the table of honour, however, was a well-kept secret up to the very last moment. The table of honour was decorated with a 22-metre-long flower garland in red, pink, white and yellow and with bowls holding peaches, pomegranates, apricots and pears. The rest of the Blue Hall was decked with garlands of flowers along the balustrades and the pillars were decorated with one-metre-high, pointed compositions of lilies in mild colours. The 25,000 flowers that were used were a gift from the Tourist Office in San Remo.

The Society of Chefs of the Year had been commissioned by the Nobel Foundation to create the menu in collaboration with seven chefs from the City Hall restaurant. The Chef the Year in 1993 was the project leader. The preliminary preparations were made as early as March and in May the ingredients were decided on. Then began the work on the menu. This resulted in ten first courses and ten main courses which were all prepared and test-eaten internally by the Society. Three first courses and three main courses were then selected. An innovation this year was that the Society should also be responsible for the dessert. The head pastry cook at Operakällaren was given the task and he, too, devoted the summer to preparing three suggestions for desserts. In September, the people involved got together and test-ate all nine suggestions. Taste, colour, serving and garnishing were discussed and when they had finished eating, the menu was decided. The Society had also been entrusted with the task of buying the ingredients. This year, there was a somewhat unusual worker in the kitchen, the Gastro-Max scholarship holder. This scholarship is awarded to one student at the Swedish schools of catering, so he or she is not a fully trained chef or cold-buffet manager. Among the serving staff were the seven oldest students at the Allé Restaurant School in Hallsberg. These were the only serving students from the schools in Sweden. The total number of waiters and waitresses was 228. The total amount of tableware used for setting the tables was 11,200 pieces of porcelain (eight per guest), 9,800 pieces of cutlery (seven per guest) and 5,600 glasses (four per guest). All in all, 360 bottles of champagne were drunk (two glasses per guest), as many bottles of wine, 180 bottles of dessert wine (two glasses per guest) and 58 glasses of brandy (one glass per guest).

The royal family were represented by the King and the Queen, Princess Lilian, Crown Princess Victoria and Prince Carl Philip but not Princess Madeleine, who was studying in England. Since the number of previous laureates at the banquet was so large, there were very few representatives of Swedish business life. All the political party leaders were invited and only the Prime Minister had sent his regrets as he was occupied by visits from the European Union. Never before had the banquet been attended by so many guests (1,406) and now the Pauli Room had also been opened at the southern end of the City Hall with tables for 92 guests. In addition, there were 200 students who had their dinner in the City Hall restaurant. Of the total number of guests, 25 were served alternative menus for various reasons.

The menu was described as one that would enchant even the most pampered of gourmets, or as being so elegant and light that it matched the ceiling of the Blue Hall, which had a summer sky with wisps of cloud projected on it. The entertainment consisted of Italian classical music by Giuseppe Verdi. The atmosphere in the Blue Hall was described as unusually lively and merry; there was no feeling of royal reverence and the guests chatted and laughed together round the tables. After the dinner and the speeches by the laureates, the ball began in the Golden Hall, where the repertoire consisted mainly of American evergreens.

Chapter 5

The Nobel Banquet Menus and the Royal Court Banquet Menus, 1901–2001

This chapter deals with the food and drink served at the Nobel Banquets on December 10th, 1901–2001. As the Swedish court began to give banquets for the laureates in 1904 and has continued ever since, the menus for these banquets are also presented here, together with a comparison of how these menus for December 10 and December 11 were similar and differed. This chapter is divided into the same ten-year periods as the previous chapter.

1901–1910

During the first decade, hors d'oeuvres were a very popular first course, being served seven times out of ten. Unfortunately, it is not possible to ascertain what kind of hors d'oeuvres they were. Another relatively common first course in the first ten years was consommé served in a cup, but again the information about what kind of consommé it was is missing. However, mock turtle soup was served the first two years, which is as often as Dover sole was served in the form of poached fillets. Variations in the first course at the royal court were not as great in the period 1901–1910, only various kinds of consommé (sorts unknown) and turtle soup. It must be remembered, however, that the court did not begin to arrange banquets for the laureates until 1904 and that there was no banquet in 1907 when Oscar II died. It is interesting to see that the first courses are comparatively similar at both banquets. All in all, 22 different first courses were served during this period, only six of which were at the court banquets.

As for the main course on December 10 during the first ten years, poultry dishes take pride of place with chicken, hazel grouse, pheasant and partridge. Couscous salad was served with the partridge, which must be considered very early in 1903. Couscous was probably not at all common in Sweden at that time but it is difficult to know for sure when the oldest written record in Sweden was since the word is not in the Swedish Academy's dictionary. Brill and turbot were only served as fillets, while pikeperch was served as a pâté and lobster represented the shellfish. Main courses in the form of vegetable dishes also occur now and then: artichoke hearts, asparagus and haricot vert. These are common first of the season vegetables in Sweden and are at their best in the spring, so it is interesting to see that they were chosen to be served during the darkest month of the year as main courses in their own right. All in all, 29 main courses were served on December 10 during the banquet's first ten years, the explanation being that there were large menus with up to seven courses. At the banquets on December 11 at the royal court, poultry also played a major role in the main courses. However, the variation is somewhat different, goose liver occurring and partridge being served as cutlets and roasted whole. The same applies to chicken. Fowl were represented by pheasant alone and the meat dishes were lamb and beef. Dover sole and salmon were served as fillets and here, too, lobster was the only shellfish. The royal court also served vegetables as a main course, but some of the dishes were not entirely vegetarian, celeriac with bone marrow, for example. Artichoke hearts were also served as a main course as well as fried vegetable with salad. A stock was also on the court menu but it is not stated whether it was

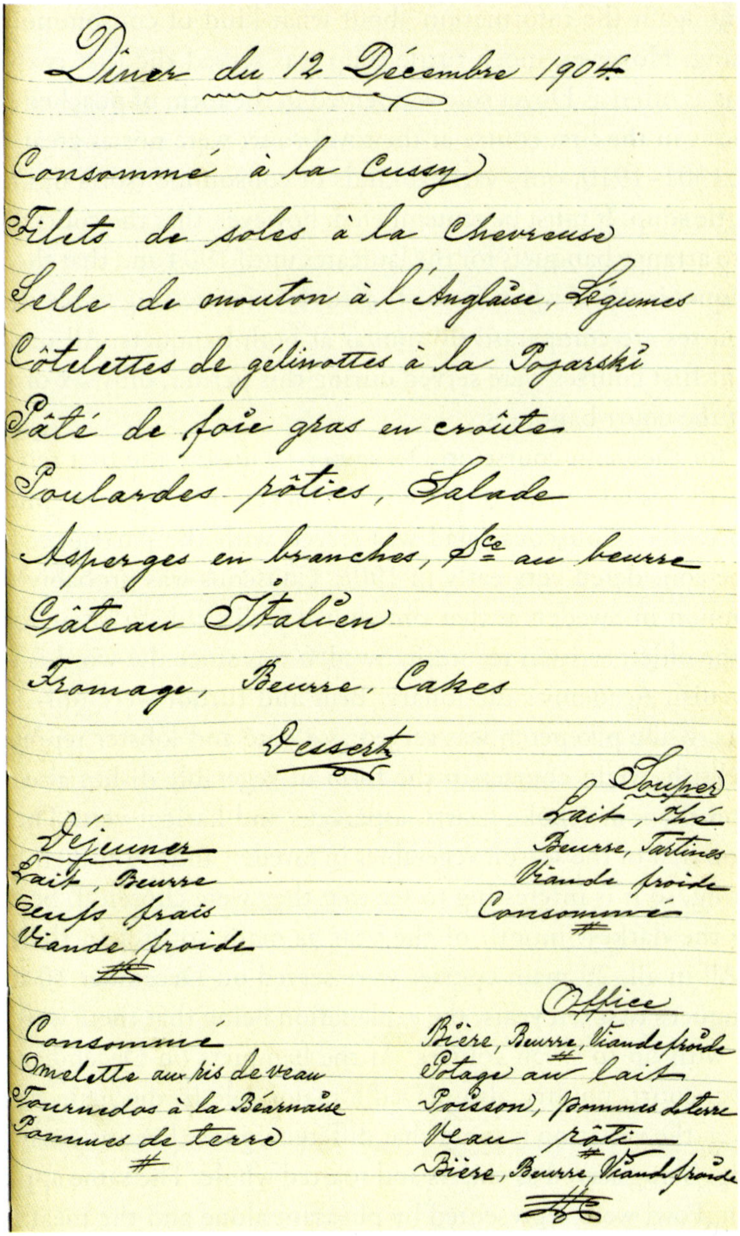

Dîner du 12 Décembre 1904

Consommé à la Cussy
Filets de soles à la Chevreuse
Selle de mouton à l'Anglaise, Légumes
Côtelettes de gélinottes à la Pojarski
Pâté de foie gras en croûte
Poulardes rôties, Salade
Asperges en branches, Sce au beurre
Gâteau Italien
Fromage, Beurre, Cakes

Dessert

Déjeuner
Lait, Beurre
Oeufs frais
Viande froide

Consommé
Omelette aux ris de veau
Tournedos à la Béarnaise
Pommes de terre

Souper
Lait, Thé
Beurre, Tartines
Viande froide
Consommé

Office
Bière, Beurre, Viande froide
Potage au lait
Poisson, Pommes de terre
Veau rôti
Bière, Beurre, Viande froide

The year 1904 was the first year the Royal Court held a banquet for the laureates. Oscar II was the King at that time. No printed menu has been preserved from that year but there is a handwritten menu that the palace chef produced. The dinner consisted of consommé, fillets of Dover sole, saddle of lamb with vegetables, pullet cutlets, goose-liver pâté baked in pastry, roast chicken with salad, asparagus and butter sauce, Italian waffles, cheese, butter and biscuits and a dessert that is not specified. © Slottsarkivet, Stockholm a vegetable stock or a meat stock. There were four fewer main courses on the court menu than on the menus for December 10.

The desserts on December 10 were of varying character, with preserved fruit and fruit tartlets dominating the picture; altogether they were served 12 times. Parfait was only served once, while various flavours of ice-cream occurred eight times. In addition, there were pastries and friandises — small, delicate morsels. The difference between parfait and ice-cream is that parfait contains more cream. Incidentally, the word parfait came into Swedish around 1900, which may explain why it occurs so seldom in this context. Ice-cream came to Sweden from France and is listed in 18th-century Swedish cookery books. The fact that the Nobel menus in this period were clearly inspired by France may explain why it occurs so often. All in all, 25 different desserts were served during the banquet's first ten years.

The desserts served on December 11 varied more than those served on December 10. A noticeable difference between the desserts on these two days is in the occurrence of fruit; on December 11, fruit was served only once. Here, too, we find small pastries or sweetmeats but also very small pastries that were not served on December 10. Another difference is that cheese, butter and biscuits were served as a dessert on December 11, which was never the case on December 10. Ice-cream was served only twice. The word Dessert also appears on the menu, which is a little surprising as in the financial accounts ice-cream is only listed as dessert. At the banquets where ice-cream was served, it came after the fruit. What it consisted of, however, is shrouded in darkness since there is no information available. Altogether, 16 different desserts were served.

A few drinks stand out among those served at the banquets on December 10, namely G-H Mumm Cremant and Porto Sandeman, which were served five and six times, respectively. Throughout this period, either sherry or a sweet port were served with the soup, whatever kind of soup it was. Today, a creamy mushroom soup is not associated with sherry or port, but in those days it was not at all unusual to drink sherry or port with soups, so here we can see a traditional choice of drinks for the various courses. Another very obvious feature during the first ten years is that the same dry champagne was served with the dessert as the banquet began with. All in all, 18 different alcoholic drinks were served during the first ten years, but then it must be remembered that

1907 is not included since the banquet that year was cancelled owing to the death of Oscar II. As for drinks served on December 11 during this period, we have unfortunately no information at all.

1911–1920

The second decade in the history of the Nobel Banquets was overclouded by the First World War, which meant that no banquets were held between 1914 and 1919. The first held after the war came in 1920, when there were two, one in June at Hasselbacken and one in December at the Grand Hotel. Unfortunately, the menu for the banquet in December has been lost, so we know nothing about what was served there. Nor did the newspapers print the menu. In the years the banquets were held during this period, there were only two first different courses, consommé and mock turtle soup. Hors d'oeuvre, which was frequently served in the first ten years, did not occur at all. Instead, the first courses are more reminiscent of the first courses served on December 11 in the previous period. The court banquets, too, are dominated by consommés. There was a little more variation among the first courses, however, as both croustades and vol-au-vents were served, which were not on the December 10 menus.

Among the main courses on December 10, cold salmon, brill and fillets of turbot were the fish dishes served. Quails represented the fowl, but pullets were also served. Saddle of venison with artichoke hearts, cauliflower and béarnaise sauce was also clearly listed on the menu. It is unclear what exact kind of meat was hidden behind the terms saddle with béarnaise sauce and fowl, but they indicate that there was a fairly strong element of game served during these ten years. As in the previous ten years, vegetables were served alone as main courses, including both artichoke hearts and asparagus. Vegetables were also served with game dishes. Although the First Word War interrupted the festivities, altogether 16 different main courses were served. In the same category of courses at the court banquets, salmon and turbot occur among the fish dishes, but Dover sole was also on the menu. Game was not so evident as during the previous period, but pheasant baked in pastry occurred. Among the domestic poultry, chicken and, for the first time, turkey were served. Roast beef with vegetables, loin of veal and lamb medallions represented red meat. Here, too, there were purely vegetable courses like salads made with celery, tomatoes and artichoke hearts. Goose liver, partridge and lobster were not served at all during this period.

There was a great variety of desserts served on December 10. Parfait was served in three forms, praline, apricot and Nobel, and for the first time a charlotte was on the menu. What lies hidden behind the term pastries, fruit and fruit with dessert is difficult to say, as there is no information available. At the court banquets, fruit was predominant as dessert; parfait occurred twice and ice-cream only once. However, we do not know what flavours they were. The same difficulty arises about the desserts called sweetmeats, pastries and small cakes. Since the sweetmeats are separate from the other pastries and small cakes, it might be thought that they were chocolates. Cheese, butter and biscuits were not served at all.

Bearing in mind that there was a world war in progress during this ten-year period and that the festivities then were all cancelled, it is remarkable how many dishes were served at the four banquets on both December 10 and December 11. Even though there were only two different first courses on both dates, a total of 16 different main courses on December 10 and 19 different main courses on December 11 were served; the number of desserts on December 10 was seven and on December 11, eleven. This was because comparatively large menus were provided during this period, as in the previous period, despite the world war. The total number of first courses was 23, of main courses, 35 and desserts, 18.

The drinks at the Nobel Banquets follow the same pattern as in the first decade, but there was a greater variety of brands. There were several different kinds of sherry, port and dry champagne which were only served once, like Charles Heidsieck sec 1900 and Chateau Lafite from the same year. During these ten years, the same sort of dry champagne was chosen to be served with both the dessert and as a pre-dinner drink. As we have seen earlier in this chapter, the desserts are rather sweet in character and today we should not choose to serve dry champagne with a sweet dessert but a sweeter drink. During this decade, 14 different wines were served on December 10 which is only four less than in the previous period, but we must remember that no banquets were held between 1914 and 1919. Thus there was an increase in the number of wines during this period compared with the previous period. As for drinks on December 11, we have, as before, no information.

1921–1930

In the period 1921–1930, the same trends were followed for the first courses as in the previous period, that is, there were only two variants, turtle soup and various kinds of consommé. The consommés were mostly based on veal and were served with cheese pâtés, sherry and truffles. There was also chicken consommé served with tongue and truffles, a game consommé with croutons and a consommé which is only described as clear. Two years are missing from this presentation, 1923 and 1924; the 1923 menu has disappeared, and in 1924 none of the laureates were able to come to Stockholm, so no banquet was held. An innovation for the first courses at the court banquets on December 11 was caviar, but it is not clear which kind it was; not too wild a guess would be either Russian or Iranian. Here, too, consommés were predominant. as on December 10. However, it is not possible to find out what sorts of consommé they were except that thick soup was served once. There was no court banquet in 1924 for the same reason as for the Nobel banquet.

Fish dishes formed the majority of the main courses on December 10. Turbot occurs several times, served with several kinds of garnishing like oyster, truffles and white wine sauce. Salmon was served with truffles, coxscomb and mushrooms while brill was served with tomatoes and mushrooms. Black grouse and hazel grouse represented fowl, the former served with pâté filled with asparagus and truffles while the latter was served with artichoke hearts, chicken mousseline and Madeira sauce. Duck was served with artichoke hearts and orange together with bacon, onion, peas and lettuce while turkey was served with lettuce, jelly and artichoke hearts. Lamb was the only domestic animal on the menu during this period, accompanied by spring vegetables, artichoke hearts, various vegetables and Sauce Choron. The difference between spring vegetables and vegetables is rather unclear. In general, spring vegetables are the first vegetables of the season (interestingly enough, these were served in December and were probably imported). A favourite vegetable as garnishing for meats during this period seems to have been artichoke hearts, which were served with both poultry and lamb. Generally speaking, during this period the menu consisted of two main courses, one of which was fish and the other meat (poultry or lamb).

As for the main courses served on December 11 between 1921 and 1930, the variation was greater than on December 10. Pheasant was still the most frequently served fowl, but other birds like chicken and partridge were also served. However, pheasant was not the only game as both reindeer and venison were served, both roasted. Beef and pork represented domestic cattle. The marine dishes were salmon, Dover sole, turbot and lobster. In contrast to December 10, several vegetable dishes were served as main courses. One of these was not entirely vegetarian since it consisted of pea purée with diced pork. The other "green" dishes were celery with croutons, artichoke hearts with asparagus, green salad, vegetable stew and vegetables alone. What the vegetables were is shrouded in darkness since there is no written information available. For the first time in the 30 years of the banquet's history, omelette was served as a main course. The total number of main courses on December 11 during this period was 34, more than twice as many as for December 10.

As before, cold desserts were the favourites for December 10, ice-cream soufflés, bombes glacées and parfaits. The latter were served with fruit cocktail and had a centre of apricot ice-cream. Fruits like pears and pineapples were also served with ice-cream and garnishing like raspberry sauce and maraschino. Fruit was predominant among the dessert served on December 11, occurring ten times, but it is not clear which fruits were served as it says "fruits of the season" in the financial accounts. Whether they were Swedish or imported fruits is not stated. This might suggest that the other "fruits of the season" were in fact Swedish. Ice-cream is the most usual dessert here, too, occurring seven times. Once again the term "Dessert" is used, without revealing what it refers to. Small pastries and cakes were also served.

In the years 1921–1930 too, most of the drinks were served only once, but there are a few winners: Madeira Old and Imperial Ruby, which were served six and four times respectively. The same trend continued in this period as in the two previous periods, that is, many different kinds of sherry, madeira and port. All in all during this period, 26 different kinds of alcoholic drinks were served. The increase in the number of drinks applied mostly to sherry, madeira and port, but there was some increase in wines, too, for example Liebfraumilch Spätlese.

Whether the missing menu for 1920 affects the number of drinks is open to question, but it is possible that the drinks served then were of the same brand as those that occurred in the other years. As for the drinks served on December 11, there is as little information as during the first 20 years.

1931–1940

During the period 1931–1940, one first course appeared that had not been on the menu before: a kind of purée soup made from mushrooms. Turtle soup and consommés were served as before, consisting this time of game consommé with rice and cheese croutons, consommé with cheese pies, veal consommé (also with cheese pies) and a chicken consommé served with truffles and ox tongue. In contrast to the previous period, consommé had to take more of a back seat among the first-course dishes in favour of caviar on December 11. Creamy soups like lobster soup were also served. The number of different first courses served on December 11 was 16, which is twice the number for December 10.

The main courses on December 10 were dominated by seafood, with Dover sole as the most frequent dish, six times during the period. It was served five times as fillets with garnishing like brandy and shrimp sauce, truffles, mussels and prawn tails, Bercy sauce and caviar sauce, and once as roulades with Scallop Saint-Jacques, mangold and champagne sauce. Turbot was poached in white wine and served with mussels, truffles and crayfish tails and patties filled with mushrooms. Brill was also combined as fillets with mussels, scampi and white-wine sauce. The only game animal on the menu was roe deer, but only once in the form of saddle with balls of game-bird forcemeat, Madeira sauce and potato cake (the first time potato was on the menu). Among poultry, duck was served most often, served with risotto-filled tomatoes, Madeira sauce, artichoke hearts, glazed chestnuts and celery. Other poultry was chicken with artichoke hearts, truffles and salad, but fowl like hazel grouse and pheasant were also on the menu, combined with fried bananas, Waldorf salad, bacon, goose liver, sauerkraut and mashed potatoes. Hazel grouse, pheasant and roe deer represented the game during the period. During the period 1931–1940, two main courses were served, one of seafood and the other of meat, mainly various kinds of poultry. Game was more evident at the December 11 banquets than at the December 10 banquets. Again, pheasant was the most frequently served dish, but reindeer,

Programme de Musique

le 11 décembre 1933.

1. *Marche "Lorraine"* . . . . GANNE
2. *Ouverture "Banditenstreiche"* . . VON SUPPÉ
3. *"Elfentanz", valse* . . . . LEHÁR
4. *La Sérénade* . . . . TOSTI
5. *Fantaisie "Carmen"* . . . . BIZET
6. *"Moment musical"* . . . . SCHUBERT
7. *Potpourri "Die lustige Wittwe"* . LEHÁR
8. *Marche du premier Régiment de la Garde Svea* . . . . . . KÖRNER

The music programme for the Court's Nobel Banquet in 1933. © Slottsarkivet, Stockholm red deer and roe deer were also on the menu. Seafood was represented by pikeperch, turbot and Dover sole. We also see an independent vegetable course again, chestnuts. It is not known how they were prepared, but it is conceivable that they were roasted and eaten with salt and melted butter.

As for desserts, various kinds of ice-cream completely dominated the picture. The soufflés were made with Curacao flavoured with violets and diced fruit. Regardless of their flavour, they were served with petits fours. Bombe glacée was served once with sugared chestnuts and petits fours. Glace au four was also served but there is no information as to the flavour or what accompanied it. This is the first time that ice-cream was the dessert for all the December 10 banquets.

Again, fruit was the predominant dessert at the December 11 banquets and as in previous decades it was listed as "fruits of the season", but pears were evidently very popular. We also see again the category "Dessert" on the menu, without any explanation of what it consisted of. As before, it came last on the menu. It might have been cake of some kind. Some ice-cream was served in the form of soufflés and parfaits but not at all as frequently as on December 10. What lies hidden behind the category "Sweetmeats" is unknown.

The pattern for the choice of drinks at the December 10 banquets is the same as for the previous 30 years, with great variations in the port, sherry and madeira. The total number of different alcoholic drinks in this period is 22, four fewer than between 1921 and 1930. However, it should be remembered that there was no banquet in 1924 since none of the laureates could come to Stockholm. Furthermore, there were the first years of the Second World War when no banquets were held, which means that 22 different alcoholic drinks were served during a period of seven years. For the first time there is information about what was drunk at the December 11 banquets. This information begins in 1932, and 1939–1940 are absent because of the war. This means that there is information about what was served at the royal court for seven years. Altogether, ten different alcoholic drinks were served and the wine list looks rather different from the list for December 10. Cognac was preferred at the court, as well as various brands of champagne and liqueurs. A bottle of Ch. Lafite Rothschild cost 5 kronor and 30 öre in 1932 and a Champagne Pol Roger Brut cost only 20 öre more per bottle the same year. Apart from the wines served at dinner, there are also the drinks served during the evening, which consisted almost entirely of Cederlunds Punsch and Whisky Old Adam, which could be mixed with Club Soda and Ramlösa mineral water. If these drinks are included, the grand total is 21. The musicians were restricted throughout the evening to drinking Punsch, Pilsner and Club Soda.

1941–1950

The first banquet held during 1941–1950 was in 1945 on account of the Second World War, so six banquets were held during this ten-year period. The first courses were soups, creamy mushroom soup, shellfish soup and veal consommé with cheese straws, and asparagus purée and small sandwiches. It is not known what kind of sandwiches they were. The first courses on December 11 were also mainly soups and the previously so frequently served consommés occur only once. Although there were six banquets, only five first courses were served on December 11 because in 1950 only refreshments were served in Karl XI's Gallery at the royal palace. All the other food was served at the National Art Museum in the form of a buffet. What this consisted of is not mentioned in the source material. This was the first time during a ten-year period when there were fewer first courses, one less, served on December 11 than on December 10.

Fillet of beef with mushrooms, truffles, artichoke hearts and Béarnaise sauce, fillet of reindeer with celery purée, and saddle of reindeer with buttered haricots verts and lingonberry cream sauce represented the red meat dishes among the main courses on December 10. The poultry included salted goose with red cabbage and horseradish sauce, while the chicken was roasted and served with celery, May turnips, onions and bacon. During this period, and for the first time, only one main course was served, so six different main courses were on the menus on December 10 between 1945 and 1950.

Once again we see a predominance of pheasant as the main course on December 11. It was served, for example, as slices of pheasant breast covered with goose liver, port jelly and cherries. The other times it was probably served roasted whole since only the word pheasant is mentioned. Dover sole was on the menu as fillets with lobster sauce, and turbot, also filleted, but with unknown garnishing. For the first time, we see Prague ham with champagne jelly stuffed with paprika. Turkey was roasted but there is no information about the garnishing. Game was represented by fillet of moose, which probably came from a royal estate, but also saddle of reindeer with game sauce and finally venison, both roasted and as a saddle. Altogether, at the five banquets on December 11 between 1941 and 1950, 15 different main courses were served, which is twice the number served on December 10 in the same period although there was one more banquet on December 10.

There was a great variety of desserts served on December 10, 1941–1950. Only two of them were based on ice-cream, one described as Bombe Glacée à la Stadshuset, but what that means is unclear, and the other Nobel Glace Parfait served with petits fours. Fruit is frequently on the menu in various forms like preserved pears with ice-cream, Poire Belle Hélène and apple pie with vanilla sauce. Napoleon tart (millefeuille) was also served as a dessert on its own. This richness of variation contrasts strongly with the previous decade, when only ice-cream was served. In contrast to the desserts on December 10, ice-cream in various forms dominated the desserts on December 11. On one occasion, ice-cream was served with candy floss in the shape of a Rococo lady. After ice-cream come various fruit desserts, on one occasion candied, otherwise fresh fruit. Why oranges should have been distinguished from the other fruit is somewhat of a mystery, but they were served as a dessert in their own right. Chocolate pralines were also on the menu as distinct from other sweets that were served. What the small cakes consisted of is not stated in the source material.

All in all, at both banquets between 1941 and 1950, 11 different first courses and 21 different main courses and desserts were served despite the Second World War It is remarkable that the menu on December 10 was reduced so that only three-course dinners were served. At the royal palace more than twice as many main courses were served as on December 10. During this period, too, we see for the first time menus for December 11 printed in the newspapers. Until 1945, they only mentioned that the King and Queen had held a banquet and who the guests were, but not a word about the menus.

During 1941–1950, 12 different alcoholic drinks were served, but we must remember of course that the Second World War prevented any banquets from being held between 1941 and 1944. The only note

The Court's Nobel Banquet on 11 December 1949 in Vita Havet, Stockholm Palace. This is the oldest picture I have found of the Court's banquet for the laureates. © SVT/Bild

I have been able to find about wine on December 11 is for 1948, when Champagne Pol Roger 1934, Ch. Haut Brion 1929, Dry Amontillado Sherry and Barats Port were drunk. We may presume that other sorts were also served during the evening.

1951–1960

The period 1951–1960 is the first period after the Second World War when the banquets were held without a break, even though those in 1956 were rather reduced on account of the situation in the world. On 24 November, 1956 it was stated that the Nobel festivities would be modest because of the critical international situation and that they would mainly focus on the prize-awarding ceremony at the Concert Hall. The prizes were presented as usual while the banquet for the laureates was held at the Stock Exchange. 18th-century music was played during the dinner, which was followed by the traditional students' tribute to the laureates. The event was described as an intimate Nobel Banquet.

Salmon was served on most occasions in the form of pirogues with brown butter, smoked with spinach cooked in cream, poached in red-wine sauce and fried with sauce verte. Lobster was served with asparagus, truffles and Béarnaise sauce and as a soup. Fillets of Dover sole poached in red wine with bacon and mushrooms, and fillets of turbot with champagne sauce and buttered rice were the seafood, which dominated the first course. Consommé was served as well, but it is uncertain what consommé it was more than that it was clear. In comparison with December 10, clear consommé was the most frequently served first course at the palace banquets. Unfortunately, the source material does not tell us what sorts of consommés they were. Creamy soup was also on the menu in the form of chicken soup with asparagus tips and half-dry sherry, as well as a clear soup that is not described in detail. In my 10-year period division, this is the first one where the number of first courses on December 11 was as large as on December 10, that is, ten.

There was great variation among the main courses on December 10. Fillet of beef was served either with beef marrow and potatoes Parisennes or with artichoke hearts, truffles and mushrooms. Fillet of veal was served with Polignac sauce and truffles while lamb occurred as saddle with pimiento sauce and green salad as garnishing. There was also saddle of roe deer with pepper sauce and chestnut purée. This is the first time young cockerel was served as the main course with Madeira sauce and salad. Otherwise it was popular to serve poultry cold: turkey (probably turkey breast) with jelly and mimosa salad, duck in port wine jelly with orange sauce and chicken with jelly and Waldorf salad. The only fowl that was served hot was roast pheasant with port wine sauce and Romaine salad. A popular garnishing for the main course during this period was various kinds of fungi and jelly. No seafood was served as a main course. Roe deer takes pride of place among the main courses at the palace on December 11, served exclusively as saddle with various sauces, for example pepper and garlic sauce. In comparison with December 10, seafood occurs often in the form of shellfish and fish; Dover sole was served several times, followed by salmon, turbot and lobster. The sole and the turbot were fried while the salmon was boiled. Again, ham was served as a main course, here in the form of forcemeat with endive salad. Cheese in pasties and as cheeseboard is also common, being served eight times out of ten. Thus it was popular to have separate cheese dishes as main courses.

During this period, Nobel Parfait was served with petits fours twice as dessert. Other kinds of ice-cream as the main ingredient were Glace au Four, and cloudberry parfait. Otherwise, fruit was often on the menu during this period. Pears cooked in champagne and brandy and preserved pears with pistachio ice-cream and chocolate sauce were also served. Another pear variant is flambé and this is the first time a flambé dessert appear in the source material. Whether or not it was flambé at the table is unclear but if it was, it must have been a sight not easily forgotten by the guests and an unbelievable feat of skill and timing by the waiters. Apple savarin with rum and whipped cream, fruit savarin with maraschino and fruit Bavarois with candy floss were other fruit desserts. On December 11, too, fruit was popular as dessert. Which fruit it was is difficult to say since, just as in previous decades, it only says fruit of the season or mixed fruit of the season. Desserts based on ice-cream are not far behind fruit. It appeared in several flavours such as pineapple, strawberry (decorated with royal sugar crowns), orange and wild strawberry. It also appeared twice as parfait. Sweets and small cakes are not served as frequently as in the earliest periods, only three or four times here.

11. 12. 1951

Vakth. Adj. Komm.-kapten Henning	Kabinettskammarh. Greve Lewenhaupt
Handsekr. Friherre Palmstierna	Adj. Komm.-kapten Frih. af Ugglas
Professor Caspersson	v. Ceremonimäst. Kammarh. Greve Mörner
Överintendenten Setterwall	Professor Fredga
Professor Waller	Professor Ölander
Professor Hulthén	Professor Nyberg
Professor Ahnlund	Fil. Dr. Gullberg
Professor Skottsberg	Författaren Fil. Dr. Siwertz
Professor Fries	Professor Cramér
Fil. Dr. Österling	Professor Liljestrand
Professor Tiselius	Professor Berg
Fil. Dr. Hedin	Stabschefen Gen.-löjtn. Cederschiöld
Ståthållaren Gen.-löjtn. Friis	Professor von Euler-Chelpin
Fru Bödtker	t. f. Universitetskansler Engströmer
Fil. Dr. Lagerkvist	Fru Lawrence
Hovfröken Reuterswärd	Doktor Theiler
Professor Walton	Fru Seaborg
Fru McMillan	Sir John Cockcroft
H. E. Sydafr. Unionens Minister Horn	Fru Gallagher
Fru Vallotton	H. E. Storbritanniens Ambassadör Stevens
H. E. Norges Ambassadör Bergersen	Fru Butterworth
Fru Ekeberg	H. E. Statsminister Erlander
H. K. H. Hertigen av Halland	H. K. H Hertiginnan av Västerbotten
H. M. Drottningen	**H. M. Konungen**
H. K. H. Hertigen av Södermanland	Överhovmästarinnan Frih.-an Rålamb
Fru Bergersen	H. E. Riksmarskalken Herr Ekeberg
H. E. Förenta Staternas Amb. Butterworth	Fru Stevens
Fru Horn	H. E. Schweiz' Minister Vallotton
H. E. Irlands Minister Gallagher	Lady Cockcroft
Fru Walton	Statsrådet Persson
Professor McMillan	Fru Theiler
Fru Lagerkvist	Professor Seaborg
Professor Lawrence	Tjg. Hovfröken Steuch
Hovfröken Cederström	Regeringsadvokat Bödtker
Professor Siegbahn	Förste Hovmarskalken Frih. von Essen
Överceremonimästaren Frih. de Geer	Professor Hevesy de Heves
Professor Löfstedt	Envoyé Ståhle
Professor Bergstrand	Professor Bergendal
~~Professor Westgren~~	Professor Frisk
Direktör Nobel	Professor Lindblad
Professor Wessén	Författaren Fil. Dr. Hellström
Professor Antoni	Professor Lindh
Professor Ising	Professor Theorell
Professor Granit	Författaren Martinson
Riksarkivarie Andersson	v. Ceremonimäst. Kammarh. Aminoff
Ceremonimästaren Frih. De Geer	Kammarherre Tham

Förste Hovmarskalken Rudebeck

Entré

The placing list for the Court's Nobel Banquet in 1951. The name that is crossed out had cancelled after the list had gone to the printers. © Kungl. Husgerådskammaren. Photographer: Alexis Daflos

For the first time, non-alcoholic beverages like coffee and Ramlösa mineral water were listed on the menus for December 10 during this period. Coffee appears on eight out of ten menus. This does not mean that coffee was not served previously, since we know that coffee was drunk considerably earlier than 1951 at the banquet thanks to newspaper reports (see Chapter 4). Altogether, 35 different drinks were served in the years 1951–1960, two of them non-alcoholic, making 33 alcoholic drinks. This is a considerable increase compared with the two immediately preceding periods. An increase in the amount of wine and brandy served can be seen compared with earlier. The word Buffet was written as the last item on the menu for the first time in 1953, which refers to the refreshments buffet which was opened for the ball (see Chapter 4), but it is not known what was served at the Buffet. Lanson Père et fils champagne has a rather leading position among the drinks served at the palace for the laureates on December 11 during this period. It is rather difficult to work out how many different kinds of drinks were served as there is nothing to tell us what lies behind terms like red wine, sherry and port. There might be one kind of each or more. With some reservation, it might be said that there were 11 different sorts of alcoholic drinks served on December 11 in the years 1951–1960.

1961–1970

Seafood dominates the picture in the period 1961–1970, as in previous periods, for the first course. Neither game nor domestic cattle meat was served. Even the avocado that was served as a main ingredient was accompanied by seafood, lobster. Avocado is also hidden in one of the salmon dishes, filled with this fish in a cold sauce. Salmon also occurs both smoked and braised with various garnishing. Dover sole was served as fillets in white wine with curry sauce and as stuffed roulades while turbot was served smoked with a cold sauce.

If we compare the first courses served on December 11 with those served on December 10, we see that they have a completely different character. These first courses were all liquid in one form or another. Consommés occurred most often, as poultry consommé, for example. Here, on one occasion, we meet turtle soup again, which was so frequently served in the first 20 years of the banquets and had been absent for two decades. Green pea soup with whipped cream and mushrooms was also on the menu, as was creamy shrimp soup with cheese straws.

Seafood is totally absent from the main courses on December 10, as in the immediately preceding ten-year period. Game is more frequent that domestic cattle meat. The former is represented by birds like partridge, wild duck, hazel grouse and black grouse. Reindeer represented the game animals. As for domestic animals, meat from bullock, lamb, duck and pullet was served. Black grouse and capercaillie were served cold with garnishing of truffles, goose liver, Madeira sauce and Waldorf salad. Duckling was also on the menu, also cold in jelly with the same garnishing at the black grouse. Hazel grouse was served in jelly with goose liver, and Waldorf salad, wild duck was served with garlic, Madeira sauce and salad. Waldorf salad and various sorts of jelly seem to have been popular garnishing for fowl, which were usually served cold. Chicken in jelly was also combined with asparagus vinaigrette, Madeira sauce and goose liver while partridge was served with vegetables and mimosa salad. It seems that cold mayonnaise salads and Madeira sauce were popular with poultry in this period. Lamb was in the form of saddle with pimiento sauce and green salad (as in previous decades) while beef was served as stuffed fillet with truffle sauce. Fillet of reindeer was also stuffed and was served with Geneva sauce and pommes dauphines.

Comparing the main courses served on December 11 with the main ingredients of the food served on December 10, we see that the variation between seafood and meat is greater. Game is also dominated by roe deer during the period 1961-1970 as in the previous period. It was served throughout in the form of saddle with various side dishes such as chestnut purée. The side dishes served with fillet of beef are not known. Pheasant was served roasted with croutons and mushrooms. Nor can we see what fish formed the basis of fish dishes with lobster, but Dover sole was served, for example, as roulades filled with forcemeat. It was also braised in red wine and served with lobster. Salmon occurred as quenelles in pastry, and pike as forcemeat on a bed of mushrooms. Lobster and turbot were also served but the garnishings are not known. As in the previous ten-year period, cheese was on the menu, this time in the form of pastries. It must have been popular because it was served five out of ten possible times as cheese pastries with celery. After having been absent for a couple of decades, a vegetable dish in its own right was served in the form of a salad, but we do not know what vegetables were used or whether it was a hot or a cold dish.

Fruit was still very popular, dominating the desserts on December 10. Pears occur in three different constellations such as in champagne with vanilla sauce, and as Nobel Gala Pears, though what that was we cannot say. Other fruits were peaches in Grand Marnier with flavoured cream and petits fours, and pineapple in liqueur with petits fours. Pineapple also formed the basis of a parfait which was also served with petits fours. Sorbets were made from oranges, and savarin from cloudberries served with mocha cream. Finally, vanilla ice-cream was served with chocolate sauce. Chilled desserts are in the majority on December 11 in the form of sorbet, ice-cream and parfait. Sorbets are most common, with either mandarin or arrack flavouring. The ice-cream was accompanied by almond chips. The parfait had mocha flavouring. Fruit flambé occurred once in the form of pineapple with ice-cream. Again, we find fruit on the menu, but only once. This is a striking decrease from previous periods, but as before it is not stated what sorts of fruit they were. The number of times small cakes were served as dessert also decreased from previous periods to just one occasion.

During this period, coffee shares the head of the list of beverages served on December 10 with Marie Brizard and Strega liqueurs. Other drinks high on the serving list are Pommery & Greno Brut and Courvoisier brandy. All in all, 15 different alcoholic drinks were served, which is a decrease from the previous period of 50 per cent. The range of brandies served here is not so wide. As for December 11, the information available is sporadic and unclear, but this is the first time that non-alcoholic drinks are mentioned as being served during the dinner. This does not necessarily mean that non-alcoholic drinks had not been served before; they may well have been but for some reason they were not mentioned. Alongside the non-alcoholic alternative, there was champagne, red wine, port and sherry, which were served in the previous period It is worth noting that this is the first time during the period that the wines are written out on the December 11 menu, in 1964, when the guests were served La Ina, Ch. Cheval Blanc 1950 and Rebello Valente Vintage 1943.

1971–1980

As in the previous two decades, seafood completely dominated the first courses on December 10. Salmon was popular, served several times smoked with poached egg and Hollandaise sauce and once on a bed of spinach with poached egg. Turbot was on the menu in several forms, as mousse garnished with truffles in champagne jelly, smoked with Hollandaise sauce and as cold fillets with whitefish roe and Hollandaise sauce. Brill was served cold with shrimp sauce, while Dover sole was poached in vermouth and served with pilau rice. Finally, lobster was eaten cold. Hollandaise sauce seems to have been popular with fish in this period.

No banquet was held at the royal palace in 1972 as Princess Sibylla had died. In the years when a banquet was held for the laureates at the palace, creamy soups seems to have gained popularity among the first courses while mock turtle soup was served only once. The creamy soups included cream of mussel, green pea and lobster and pilgrim scallop. There was also partridge consommé and pheasant consommé. Another variant was clear chicken soup. Thus all the first courses served on December 11 were in liquid form.

As in the two previous ten-year periods, seafood was absent from the main courses on December 10. Game occurred in the form of snow grouse, hazel grouse, roe deer and moose. Looking more closely at the poultry, chicken was served with cream sauce, Brussels sprouts, carrots and mushrooms, while snow grouse was served either roasted with truffles, goose-liver sauce, rowanberry jelly, salad of the season and potatoes or grilled with juniper berries, and with rowanberry jelly and salad. Hazel grouse was preferred cold as a mousse with truffle sauce and Waldorf salad while duck was braised in Calvados and served with apple compote. As for lamb, this was baby lamb steak with Madeira sauce and tomato salad, while beef was served as fillet. Roe deer was roasted and served with Geneva sauce and salad while veal was roast saddle with morel sauce and haricot-vert salad. Reindeer was served as fillets (like the beef) with chanterelles, aquavit sauce and Lyonnaise potatoes. Compared with the previous ten-year period, the main courses seem to have greater variation of both the main ingredients and the garnishing. Again, roe deer predominates among the main courses on December 11. It was always served as saddle, often with wild-pepper sauce. Apart from roe deer, there was no meat of either poultry, domestic cattle or game. Seafood, however, consisted of Dover sole, turbot, charr, salmon and shrimps. Fish soufflé was served, but it is not clear what sort of fish it was, only that it was served with rice; shrimp soufflé was also solved. Salmon, too, was served as soufflé baked in pastry. Dover sole, turbot and charr were served as fillets, and the two first-named with lobster and stuffed with salmon in white-wine sauce while charr was served with shredded vegetables. Again, cheese was served as a main course in its own right, this time and often as cheese balls, sometimes made from Stilton cheese.

Chilled desserts in various forms are completely dominant on December 10: soufflés, ice-cream, parfait, sorbet and Nobel Glace Parfait. The soufflés were served as an ice-cream soufflé made with Grand Marnier, and the ice-cream was served à la Suedoise (whatever that means). One parfait was made of pineapple and the other of lingonberry, the latter served with almond biscuits; the sorbet was made of oranges. It is not known what the Nobel Glace Parfait consisted of during these years, but it is safe to say that there was a different flavour each time. On December 11, too, chilled desserts were in the majority. Only once was something else served: peach flambé with almond biscuits. Once, ice-cream soufflé was made with pineapple.

When sorbet was on the menu, it was most often made with tangerines and served with almond biscuits or almond chips, although once it was made with wild strawberries and served with almond chips.

Again, we find that coffee is the beverage that was served most frequently on December 10. Another non-alcoholic drink was Ramlösa mineral water, which was on the menu only twice but was most probably served throughout the dinner. This seems likely if we look back at Chapter 4 and see what the newspapers wrote. But now there is a change in the drinks compared with previous years. Sherry, port and Madeira have given way to wines, brandy and champagne. Altogether, 18 different sorts of alcoholic drinks were served. In 1975, for the first time, the menu specifies what the ball buffet consisted of, and that was two drinks only: Gordon's gin and Bell's whisky. During the remaining years of the period, this buffet continued to serve two drinks until 1980, when it was increased to three. Gordon's gin lasted until 1978, when it was replaced by Bols Silver Top dry gin and this continued for the rest of the period. As for whisky, the following brands were served: Stewart's Finest Old Scotch, Bell's and Long John. Campari was added in 1980. What was served on December 11 is unknown, but not too bold a guess would be that they were the same as in the preceding period.

1981–1990

Once again, seafood predominates among the first courses on December 10, but snow grouse and reindeer were also on the menu. These two were the only game dishes served as first courses.

The grouse was served as a mousse with cream sauce and as a timbale on a bed of lettuce with herb sauce. The reindeer was gravad fillet with Dijon mustard sauce. Salmon, which was served most frequently, was mousse with crayfish sauce, gravad in juniper berries with mustard sauce and morel pâté, and fresh-smoked with whitefish roe. Crayfish and lobster were the shellfish represented, the former a timbale with basil sauce and the latter a truffle pâté with cress sauce. Charr was gravad and served with green-pea sauce, eel was in the form of a pâté together with Dover sole, served with trout-roe sauce.

Compared with the first courses on December 10, those served on December 11 had a somewhat different composition. The creamy soups were green-pea soup while the clear soups were various kinds of bird soup such as grouse, pigeon, pheasant and quail. Whitefish roe was served in parsley purée, but how the salmon was served we do not know. The soufflé consisted of catfish with shrimps. and fresh goose liver was also on the menu, though what the side dishes consisted of is not revealed in the source material except that it was also served in the form of a pâté. In contrast to the first courses served on December 10 in this period, there were plenty of poultry and game bird dishes.

Relatively little seafood appears among the main courses on December 10, except for charr and Dover sole. Game is represented by moose, hare, pheasant, red deer and wild duck. The moose was as fillets with morel sauce, Brussels sprouts and jelly with leek potatoes and as marinated back with lingonberry cream sauce. The hare was served as saddle with vegetable-filled apple rings and Calvados sauce, while the red deer was served as stuffed fillet with chanterelle and morel sauce. The charr was either braised in dill and cream or grilled with juniper berries and served with trimmings like asparagus, dill cream and rice. The Dover sole appears as morel-stuffed fillets with vermouth sauce and rice. As for fowl, pheasant was roasted and served with a timbale of forest mushrooms, and wild duck with blackcurrant sauce, carrot cake and salad. Finally, lamb was in the form of saddle with cognac-cream sauce and potatoes. Again, we find that the ingredients for the main courses on December 11 are somewhat different from December 10. Again, roe deer is predominant and was served in the form of saddle with various trimmings such as frozen cranberries and chestnut purée. The occurrence of seafood and variations, however, are greater than on December 10 during the same period. Charr was served as fillets but what the garnishings were we do not know. The salmon was Norwegian fiord salmon, but there is no information about how it was prepared or with what trimmings. Turbot was in the form of roulades filled with various vegetables while the prawns were fresh. Pilgrim mussels were served in soufflés, but nothing is said about the crustaceans that were included in the seafood buffet. A shrewd guess would be that it was very varied. Dover sole was cooked in Pernod sauce. Beef was served in the form of oxtail soup. Again, we see that cheese provided the basis for a main course of its own and cheese pastries were also served as in the preceding ten-year period.

There is only one category of desserts in the years 1981–1990, the now famous Nobel Glace Parfait, so this is the first period in which only this dessert was served. The source material does not tell us what flavours were used in the ice-cream or the parfait, but the same flavour was never used twice, either in the ice-cream or in the parfait. It was served with candy floss and petits fours, sometimes crowned with the now famous caramelised N and sometimes with small biscuits shaped as Ns. When the small N biscuits were served, there was one per guest, but of the caramelised N there was only one big one on the top of each ice-cream parfait.

On December 11 too, chilled desserts were in the majority. Mousse was served flavoured with Arctic raspberry or clementine. Sorbet was grapefruit, pineapple and lingonberry. Whether the mousse and sorbet were served with anything is not known. An ice-cream buffet was also on the menu, but we do not know what the ice-creams were. However, ice-cream soufflé was flavoured with wild raspberries, cloudberries and hazelnuts. Marinated oranges were also served with vanilla ice-cream and almond chips.

Coffee is still head of the league together with Ramlösa mineral water among the drinks served on December 10, when they had ten out of ten possibles. An innovation was that non-alcoholic wines appeared on the menu as alternatives. This does not necessarily mean that they were never served before, but they did not appear in the source material. This is the first time that Grådask port was on the menu. In comparison with the preceding period, there is less brandy and not such a wide range of wines. Altogether, 13 different alcoholic alternatives were served.

As for the refreshments buffet at the ball, Long John whisky, Bold Silver Top dry gin and Campari still appeared. In 1981, the newcomers were Cachucua and Nutty Solera, in 1982, Tio Pepe and in 1983 Spendrups beer. In 1984, something happened and the drinks were now The Famous Grouse (until 1987), Queen Anne (until 1986) and Carpano Punt e Mes (until 1986). The 1985 buffet was the same as in 1984 and in 1986 Gordon's Dry Gin was added. In 1987, apart from The Famous Grouse, Glenlivet Malt, Gordon's Dry Gin and Spendrups beer, there was also Hennessy VS. The following year, the buffet looked a little different, comprising Petite Liqueur from Moët & Chandon, Johnnie Walker Red and Black Label, Gordon's Dry Gin, Baccardi rum, Hennessy VS and Spendrups beer. The following year, Gordon's gin was replaced by Absolut vodka (the first time in this context), but otherwise the buffet was the same. In 1990, Gordon's Dry Gin reappeared, otherwise the buffet was the same as in 1989. As for December 11, the newspapers wrote nothing about the drinks that were served.

1991–2001

Compared with the previous three ten-year periods, seafood did not completely dominate the first courses in 1991–2001. We see ingredients like duck and vegetables like artichoke hearts, Jerusalem artichoke and nettles, but there is no sign of game animals or domestic cattle among

Sommelier Lennart Andersson making sure that the carafes of wine agree with the right serving stations for the Court's Nobel Banquet for the laureates on 11 December, 2001. © Kungl. Husgerådskammaren. Photographer: Alexis Daflos

Sommelier Lennart Andersson decanting wine for the guests at the Palace on 11 December, 1991, the 90th anniversary of the Nobel Banquet. This has been done every year there has been a banquet for the laureates. © Kungl. Husgerådskammaren. Photographer: Alexis Daflos

the first courses. The duck was served as smoked breast in roulades with mango and turnip. Lobster appeared à la daube with creamed cauliflower and whitefish roe, and with cauliflower purée and jelly. Charr was gravad on a bed of black radish with salmon roe and honey vinaigrette while salmon appeared as a terrine together with Dover sole, dill and chive, and gravad in the form of a tartare with red-pepper cream, served with garden-sorrel sauce and whitefish roe. Cod was a slightly salted Atlantic cod with garden-sorrel sauce and whitefish roe. Halibut was served as a terrine with langoustines and marinated pilgrim mussels, whitefish-roe crème fraîche and tender salad. Artichoke hearts were on the menu marinated and filled with shrimps, crayfish and fresh fennel, while Jerusalem artichoke was served once as a cream with whitefish roe and lobster and once as a pasty filled with smoked salmon and lobster. Nettles appeared as a soup with quail's egg. Soups were common as the first course on December 11. The creamy kind was made from green peas while the others had as their basis langoustines, mussels, fennel and blue mussels. Salmon was prepared and served as quenelles au gratin on a bed of leaf spinach. Thus there was plenty of crustacean food, but marinated duck-liver pâté and smoked goose breast were also on the menu.

There was no seafood whatsoever among the main courses on December 10 during 1991–2001, but there was a balance between game and domestic cattle. The game consisted of reindeer, red deer, pigeon and quail. The domestic meat was veal, lamb, chicken, duck and guinea fowl. The reindeer was in the form of fillets served with chanterelles, kohlrabi timbale, blackberry sauce, mashed potato and sliced black radish. The red deer also appeared as fillets fried in butter before being served as a stew with forest fungi, zucchini and bacon, blackberry sauce and croquette potatoes. Chicken was served as breast fried with thyme and black pepper before being placed on the plate with Jerusalem artichoke sauce, forest fungi roulades and vegetable ragout. Duck was also in the form of breast either fried with lemon and then combined with Västerbotten timbale, chanterelle and Jerusalem artichoke ragout and herb potato purée or with sea buckthorn sauce and root vegetable compote. Guinea fowl was served stuffed together with roasted root vegetables, citrus sauce and mashed almond potato. Fried breast of pigeon was served with a ragout of Karl-Johan mushrooms and leg of pigeon, sourish raspberry vinegar sauce and potato and onion compote. Quail was also served fried, with tomatoes and green asparagus. Those times

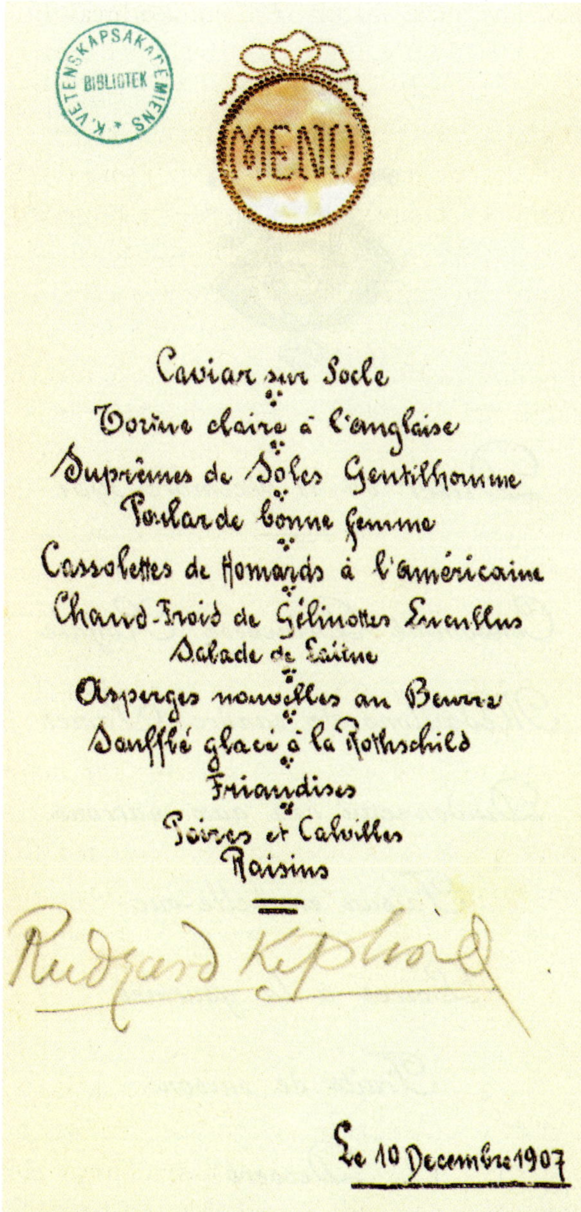
K. VETENSKAPSAKADEMIENS BIBLIOTEK

MENU

Caviar sur Socle
Tortue claire à l'anglaise
Suprêmes de Soles Gentilhomme
Poularde bonne femme
Cassolettes de Homards à l'américaine
Chaud-Froid de Gélinottes Lucullus
Salade de Saison
Asperges nouvelles au Beurre
Soufflé glacé à la Rothschild
Friandises
Poires et Calvilles
Raisins

Rudyard Kipling

Le 10 Décembre 1907

Although there was no banquet organised by the Nobel Foundation in 1907 owing to the death of Oscar II, a dinner was arranged in honour of Rudyard Kipling at the Grand Hotel, Stockholm. The menu consisted of caviar, bouillon, Dover sole, roast chicken, lobster, asparagus with butter, ice-cream soufflé, small cakes, pears, Calville apples and grapes. Unfortunately, the wines are not given on the menu. The menu is handwritten and autographed by Kipling. © Svenska Kungliga Vetenskapsakademien, Stockholm

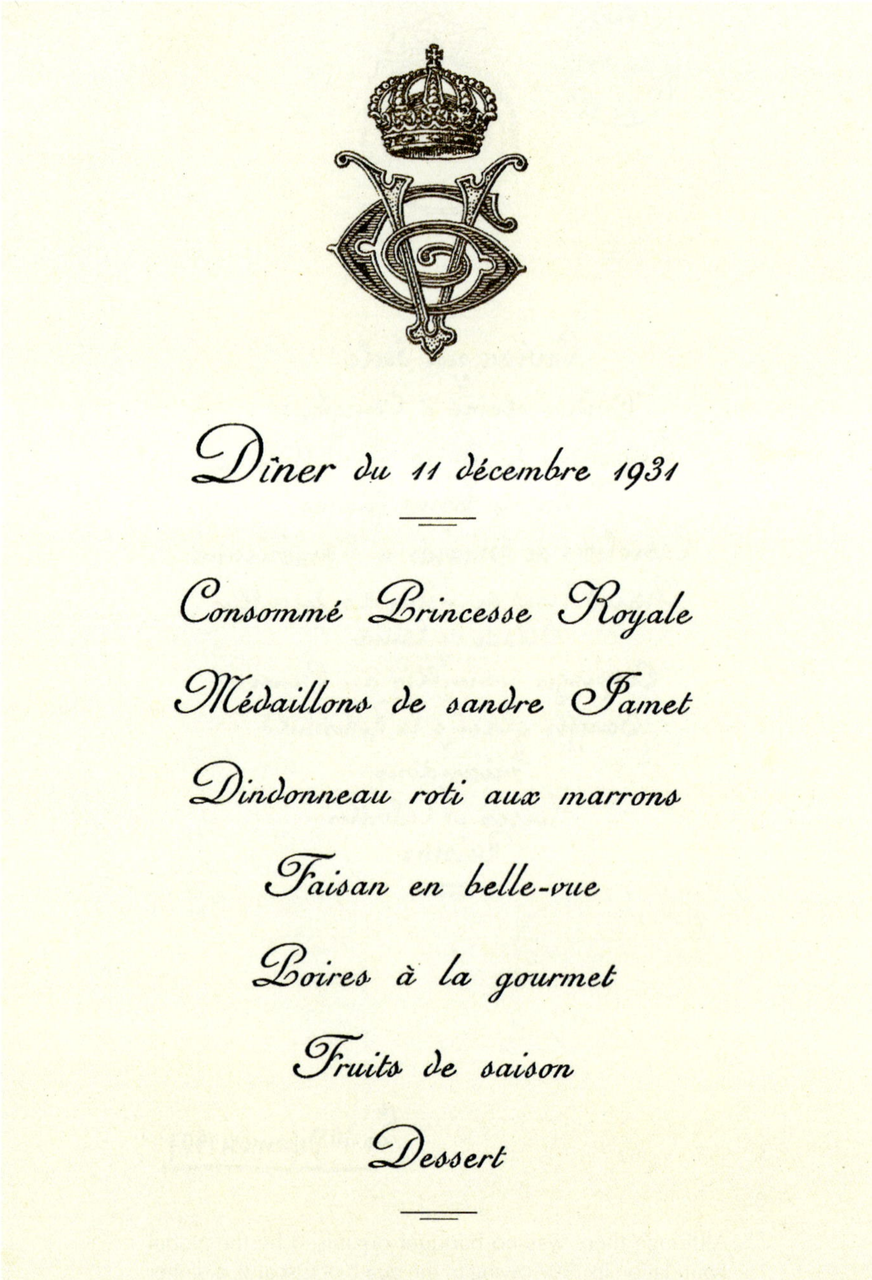

Dîner du 11 décembre 1931

Consommé Princesse Royale

Médaillons de sandre Jamet

Dindonneau roti aux marrons

Faisan en belle-vue

Poires à la gourmet

Fruits de saison

Dessert

The oldest known printed menu from the Court's Banquet on December 11 for the laureates is dated 1931 and is crowned with King Gustav V's crest. The menu included consommé, pikeperch médaillons, roast turkey with chestnuts, roast pheasant, pears, fruits of the season and dessert. What the dessert was is not stated. © Slottsarkivet, Stockholm

Dîner

du 11 décembre 1933

Caviar d'Astrakan

Consommé chartreuse

Médaillon de saumon Amiral

Quartier de chevreuil Grand-veneur

Aiguillettes de faisan Montmorency

Parfait glacé au curaçao

Fruits de saison

Mignardises

Only two years after the 1931 menu, it has changed its appearance. This menu is from the Court's Banquet for the laureates on 11 December, 1933. Gustav V is still king, but this menu is not crowned with his crest as it was in 1931. The dinner included caviar, consommé, salmon médaillons, roe deer, pheasant, parfait flavoured with Curaçao, fruits of the season and small pastries. © Slottsarkivet, Stockholm

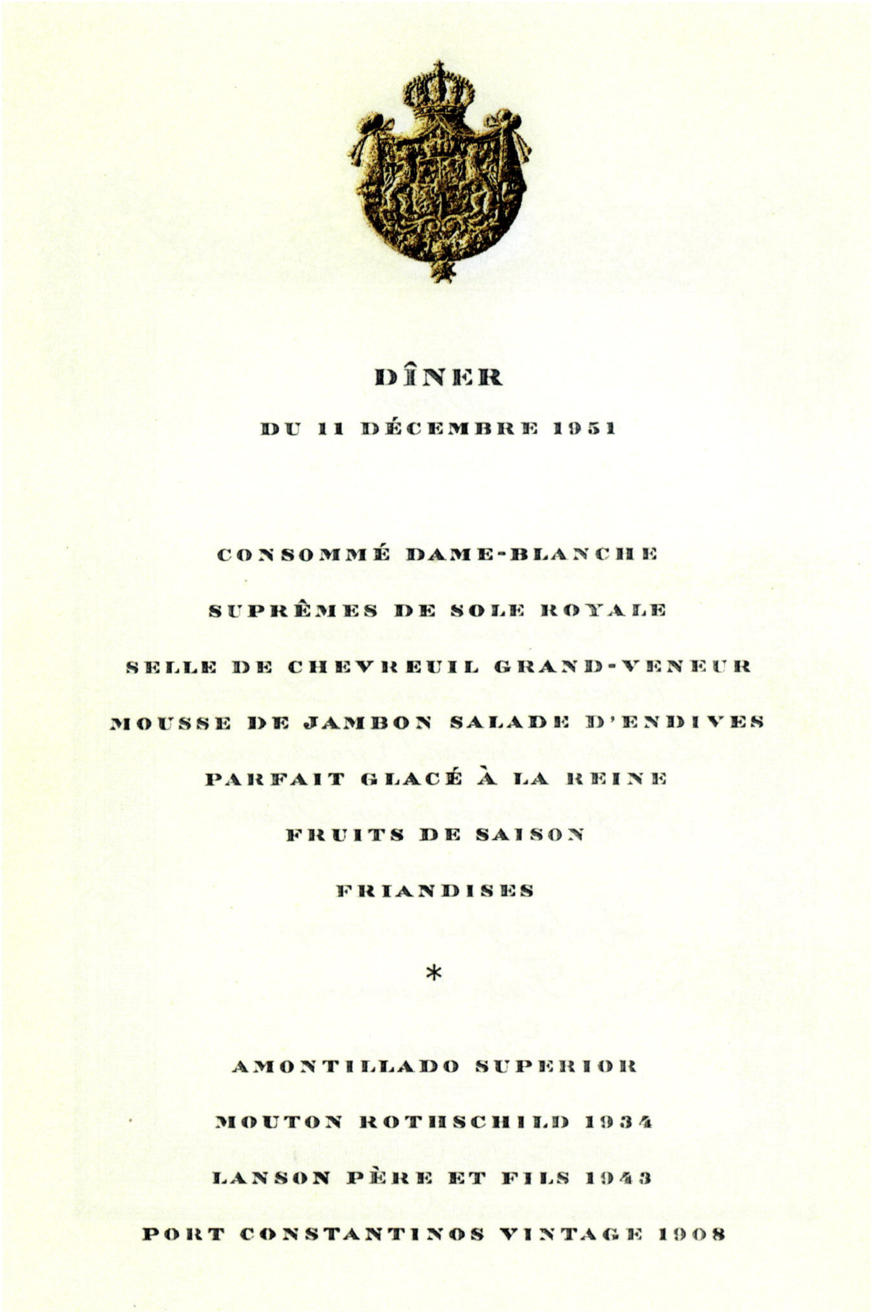

DÎNER

DU 11 DÉCEMBRE 1951

CONSOMMÉ DAME-BLANCHE

SUPRÊMES DE SOLE ROYALE

SELLE DE CHEVREUIL GRAND-VENEUR

MOUSSE DE JAMBON SALADE D'ENDIVES

PARFAIT GLACÉ À LA REINE

FRUITS DE SAISON

FRIANDISES

*

AMONTILLADO SUPERIOR

MOUTON ROTHSCHILD 1934

LANSON PÈRE ET FILS 1943

PORT CONSTANTINOS VINTAGE 1908

One of Gustav VI's early Nobel Banquets as King of Sweden. The menu is dated 11 December, 1951 and it consisted of consommé, Dover sole, roe deer, ham mousse with endive salad, parfait, fruits of the season and small pastries. Four wines accompanied the food, one of which was a vintage port from 1908. This wine had survived two world wars before it was uncorked. © Kungl. Husgerådskammaren. Photographer: Alexis Daflos

DÎNER

DU 11 DÉCEMBRE 1991

Les quenelles de saumon au gratin

La selle de chebreuil de la chasse royale
au poivre vert

La fromage Brillat Savarin

Le soufflé glacé aux framboises arctiques

★

Château Isenbourg Gewurztraminer 1983

Château de Lisse 1973

Mesimarja-Åkerbärslikör

The menu served at the Court's Nobel Banquet at the 90th anniversary on 11 December, 1991. It included salmon quenelles au gratin, saddle of roe deer with green peas, cheese and an ice-cream soufflé flavoured with strawberry as the dessert. © Kungl. Husgerådskammaren. Photographer: Alexis Daflos

DÎNER

DU 11 DÉCEMBRE 2001

Turbot poché,
Brandade de crabe et sauce à l'estragon

Filet de chevreuil de la chasse Royale

Surprise aux trois chocolats

★

Château de Rochemorin 1999

Château Lafleur-Gazin 1994

Riversaltes Vieille Reserve 1988

The menu served at the Court's Nobel Banquet at the centenary on 11 December, 2001. It included poached turbot, fillets of roe deer and three sorts of chocolate as the dessert. © Kungl. Husgerådskammaren. Photographer: Alexis Daflos

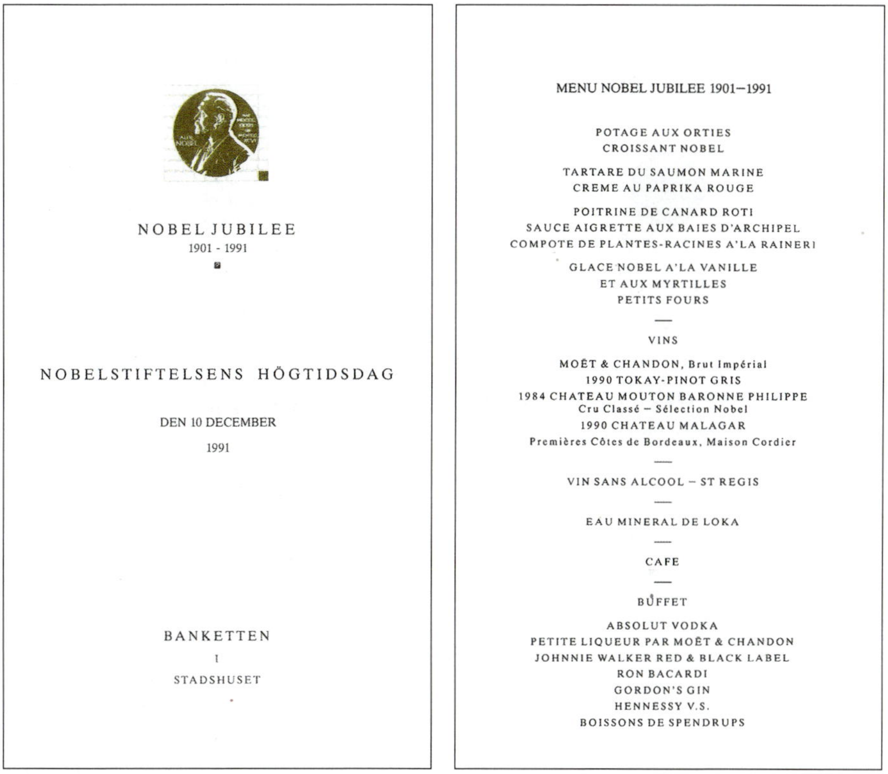

NOBEL JUBILEE
1901 - 1991

NOBELSTIFTELSENS HÖGTIDSDAG

DEN 10 DECEMBER
1991

BANKETTEN
I
STADSHUSET

MENU NOBEL JUBILEE 1901–1991

POTAGE AUX ORTIES
CROISSANT NOBEL

TARTARE DU SAUMON MARINE
CREME AU PAPRIKA ROUGE

POITRINE DE CANARD ROTI
SAUCE AIGRETTE AUX BAIES D'ARCHIPEL
COMPOTE DE PLANTES-RACINES A'LA RAINERI

GLACE NOBEL A'LA VANILLE
ET AUX MYRTILLES
PETITS FOURS

—

VINS

MOËT & CHANDON, Brut Impérial
1990 TOKAY-PINOT GRIS
1984 CHATEAU MOUTON BARONNE PHILIPPE
Cru Classé – Sélection Nobel
1990 CHATEAU MALAGAR
Premières Côtes de Bordeaux, Maison Cordier

—

VIN SANS ALCOOL – ST REGIS

—

EAU MINERAL DE LOKA

—

CAFE

—

BUFFET

ABSOLUT VODKA
PETITE LIQUEUR PAR MOËT & CHANDON
JOHNNIE WALKER RED & BLACK LABEL
RON BACARDI
GORDON'S GIN
HENNESSY V.S.
BOISSONS DE SPENDRUPS

The Nobel Banquet menu for the 90th anniversary in 1991. It consisted of nettle soup with quail's egg, tartare of gravad salmon with red-pepper cream, roast breast of duck with sea buckthorn sauce and vegetable compote, and Nobel Gala ice-cream parfait. © The Nobel Foundation

NOBEL PRIZE
CENTENNIAL
1901–2001

■

NOBELSTIFTELSENS

HÖGTIDSDAG

DEN 10 DECEMBER 2001

■

BANKETTEN

STADSHUSET

MENU

HOMARD SUR PUREE DE CHOU-FLEUR
ET GELEE DE LANGOUSTINES
SALADE DE SALICORNES

■

CAILLE FARCIE AU FOIE GRAS ET
SA POELE DES CEPES ET TOMATES SECHEES
ASPERGES VERTES ET PUREE DE CERFEUIL

■

DUO DES GLACE VANILLE ET
PARFAIT CASSIS MERINGUE

VINS

1989 LOUISE POMMERY
Champagne

1997 CHATEAU PALMER
Margaux

1998 BERNKASTELER BADSTUBE RIESLING EISWEIN
Mosel-Saar-Ruwer

CAFE

GRÖNSTEDTS RARISSIME
Grande Champagne

COINTREAU

EAU MINERALE RAMLÖSA

■

MENU Föreningen Årets Kock

The Nobel Banquet menu for the centenary celebration in 2001. It consisted of lobster on cauliflower purée with langoustine jelly and salicornia salad, Nobel roll, quail stuffed with duck liver with a ragout of Karl-Johan mushrooms, sun-dried tomatoes, fresh green asparagus, Madeira gravy and chervil purée, and vanilla ice-cream and blackcurrant parfait on a thin meringue case with thin toffee biscuits. © The Nobel Foundation

Brewing coffee for the guests on 11 December, 1991 in the Palace kitchen, Stockholm. The lady on the far left comes in every year to do this job. The coffee is brewed in the large copper kettles and then transferred to smaller pots from which the waiters and waitresses serve the guests. © Kungl. Husgerådskammaren. Photographer: Alexis Daflos

The cheese and bread are put out just before serving on 11 December, 1991.
© Kungl. Husgerådskammaren. Photographer: Alexis Daflos

Detail picture of the tables laid for the Court's Nobel Banquet on 11 December, 1996. © Kungl. Husgerådskammaren. Photographer: Alexis Daflos

The roe deer which is always on the menu of banquets on December 11. Here the dish is presented in its 2001 model. The reason why roe deer is so frequently served is that the present king, His Majesty Carl XVI Gustav, fells the deer himself. That is also the reason why roe deer is not served at the banquets on December 10. © Kungl. Husgeråskammaren. Photographer: Alexis Daflos

A really great "behind-the-scenes" picture. Here the finishing touches are being given to the dessert on 11 December, 2001. © Kungl. Husgerådskammaren. Photographer: Alexis Daflos

The table laid for the guests on 11 December, 2001 for the centenary celebration at the Royal Palace, Stockholm. © Kungl. Husgerådskammaren. Photographer: Alexis Daflos

lamb was on the menu, it was roasted fillets served either with honey-glazed carrots, black radish, fried forest fungi and apple-cider sauce or wrapped in white cabbage together with sweet-sour dill gravy, Cheddar and potato purée and small vegetables. Finally, veal was in the form of fillets fried in sage before being placed on the plate with PomPom mushrooms, tomatoes stuffed with spinach and ginger-flavoured potato cakes. Between 1991 and 2001, it was popular to serve fillets of meat, and for birds, breast. Whether birds or beasts, the trimmings were almost exclusively vegetables of various kinds.

During 1991–2001, too, roe deer dominated the main courses on December 11. Again, we find that is it almost always served as saddle, though once it was in the form of fillets with red-wine sauce and potato terrine. It was also accompanied by green-pepper sauce, rich game sauce with goose liver, winter vegetables, hunter's sauce, fresh figs and chestnut purée. Cod was in the form of fillets served wrapped in white-beet leaves while the langoustine was served au naturel. Dover sole was served as fillets with langoustine while halibut was steamed and served with sourish sorbet. Witch was served as roulades with crayfish sauce. Under the term crustaceans, we find a saffron-spiced fish bouillon whose ingredients were either pilgrim mussels, langoustines and blue mussels or langoustines in white-wine sauce with sea-urchin roe. Lobster was served with cider sauce and turbot either as fillets with white sauce and salmon roe baked with Karl-Johan mushrooms or steamed with tarragon sauce. Thus seafood played a relatively large role. In addition to these main courses, there were others which had chilled sauces like lemon-grass sorbet and Absolut Mandarin sorbet. After having been absent for a while, vegetable main courses in the form of roasted chestnuts, black radish and jelly-filled apples were again on the menu. There were also various cheese dishes like creamy white-mould cheese, Cheddar, cheese from Savoy and aromatic cheese from Burgundy.

Ice-cream totally dominated the desserts on December 10 between 1991 and 2001. Nobel Glace Parfait occurs seven out of eleven possible times, but the source material does not saw what flavours were served. The difference between this and Nobel Glace is rather unclear but possibly Nobel Glace did not contain any parfait, only ice-cream. However, both sorts were served with candy floss. Spiced ice-cream contained pineapple sorbet and was served with a compote made of pineapple and passion fruit while vanilla ice-cream contained lingonberry sorbet. There was also an ice-cream variety called "Two sorts of ice-cream", but we do not know what that comprised.

December 11 also had many chilled desserts: soufflé made of Arctic raspberries, and fresh pineapple flavoured with Maraschino and exotic fruits; and sorbet flavoured with pineapple, served with Arctic raspberry pastries. The parfait was made of mocha and served with giant coffee beans. As for ice-cream, it was sometimes flavoured with saffron served with mulled wine sauce, sometimes oven-baked and served with wild raspberries and sometimes made with cloudberries. Chocolate was only served once as a dessert, with three different kinds. Fresh fruit occurred five times as a dessert in its own right, but we do not know what the fruits were.

Two important anniversaries took place in the period 1991–2001, the 90th and the 100th anniversary of the prize-awarding ceremony. This is reflected in the number of alcoholic drinks that were served, as many as 35. As for the non-alcoholic drinks, three different types of mineral water were served, though Ramlösa topped the list with nine times compared with once each for Loka and Carl von Linné. An innovation was to serve Tokay wines and Swedish punsch. Brandy does not occur so frequently while there were more brands of champagne and wine.

The refreshments buffet at the ball in 1991 was not much larger than in 1990 although it was a grand 90th-anniversary year. In 1991, it consisted of Absolut Vodka, Petite Liqueur par Moët & Chandon, Johnnie Walker Red and Black Label, Bacardi rum, Gordon's gin Hennessy VS and Spendrups beer. The following year, Hennessy VS was replaced by Hennessy VSOP Privilège, otherwise the same drinks were served, as they were also in 1993. In 1994, beer was taken away and Bacardi rum was replaced by Bacardi Premium Black and Calvados Boulard Hors d'Age. The following year, Drambuie, Finlandia vodka and Lagavulin Single Malt, 16 years old were added and beer was back. The same pattern was repeated in 1996. The 1997 buffet looked somewhat different, comprising Calvados Boulard, Drambuie, Campari Bitter and Bell's 8-year-old whisky besides Absolut Vodka, Hennessy VSOP Privilège, Gordon's gin, Johnnie Walker Black Label and Spendrups beer. For some reason, the contents of the refreshments buffet are not given after 1997 and the word Buffet was also removed. Information about what was drunk on December 11 is as meagre as before.

The total number of courses served in the period 1901–2001, including the two centenary years, was: first courses, 202; main courses, 386; desserts, 272, making a grand total of 860 dishes.

It should also be remembered that there were no banquets in certain years. If the years when there were no banquets on December 10 are deducted, there were 88 banquets (the years of the world wars plus 1924 are excluded). It should also be remembered that two menus are missing, for December 10, 1920 and December 10, 1923. Thus the number of courses should be somewhat higher than they are now. If we then count how many years the royal court held its banquets, we reach a total of 83 (here the years 1901–1903, 1907, 1924 and the war years are excluded). Having discounted these years, there is only one word to describe the total number of dishes served, and that is: IMPRESSIVE!

The source material for the festivities on December 11 is not reliable enough to allow a survey of the alcoholic drinks that were served. Concerning the banquets on December 10, certain calculations can be made. What are missing from the menus are the alternative alcoholic drinks such as kosher wines. From newspaper articles we know that such wines were served. The ten-year period when most wines were served is 1951–1960, with 33 different sorts. Next comes 1991–2001 with 31 different sorts, then 1921–1930 with 26 sorts and finally, 1931–1940 with 22 different sorts. It is interesting that the peak period is not 1991–2001, when two great anniversaries were celebrated. The period when fewest sorts were served is 1941–1950, with 12 different sorts of alcoholic drinks, probably due to the Second World War. The years 1981–1990 only have 13 different sorts, which cannot be explained by a war probably but most likely because it was decided to concentrate the range of drinks. The refreshment-buffet drinks served during the ball are not included since the guests themselves paid for them.

Chapter 6

Good Food — Good Conversation

This book has described an event that came about thanks only to the fact that a man named Alfred Nobel, "after mature consideration" established in his last will and testament a fund with the capital that was left after he had bequeathed some of it to his closest heirs and servants. The interest on the capital in the fund was to be distributed in the form of a prize divided into five categories: physics, chemistry, medicine, literature and peace. These prizes were to be called the Nobel Prizes and were to be presented on the day of the donator's death, December 10. The first prizes were presented in 1901, five years after the donator had died. Ever since the beginning, a banquet has been held which is generally known as the Nobel Banquet. In 1968, another prize was created by the Bank of Sweden which was first presented in 1969 and has since then been presented at the same ceremony as the other prizes. This new prize is generally known as the Nobel Prize for Economics, but its proper name is the Bank of Sweden Prize in Economic Sciences in Memory of Alfred Nobel.

I believe that there is something that can be called "Gastronomic Man". This gastronomic man is complex, differing from person to person, but it is possible to see certain common features. These consist of a number of factors such as mankind's basic needs, edibility, accessibility, the five senses, the relationship between gastronomy and other sciences and gastronomy as a science in its own right. The latter factor comprises food culture, religious beliefs, mealtimes, table manners, table placing, tableware and how meals are experiences by a guest in terms of colour and lighting, sound, design and the contents of the menu and its layout. All these factors must be taken into consideration when arranging an event such as the Nobel Banquet. In addition, there is the fact that Nobel prize-winners very seldom come from the same nation.

During the period 1901–2001 there were 48 nations that received a Nobel Prize. This means that the guests at a Nobel Banquet come from many different gastronomic spheres and from widespread geographical areas. Together, they represent many fine gastronomic cultures and very few are used to the special features of Scandinavian food and drink.

In 1901, when the first prize-awarding ceremony was held, the banquet took place at the Grand Hotel, Stockholm and was attended only by men, 113 in number. On this first occasion, it was difficult to attract enough attention to fill the Academy of Music where the ceremony was held, so the gentlemen had to go home, dress up their servants and take them to the Academy. This seems almost unbelievable when you know that nearly 1,500 people attended the banquet in 2001. The banquet that attracted most women was undeniably in 1909, when the Swedish writer Selma Lagerlöf was awarded the Literature prize. Until 1930, the banquets were held at the Grand Hotel, and the number of guests grew every year. In the end, there were too many, so the banquet was moved to the Golden Hall at the Stockholm City Hall — a move which had been advocated for some time. Despite a very successful banquet there, it was back at the Grand Hotel the following year.

However, in 1932, it was moved permanently to the City Hall, to begin with in the Golden Hall while the students ate in the Blue Hall. During this period, the students had a different menu from that in the Golden Hall. One problem that arose was that there was no real organisation for the speeches. Anyone who wanted to give a speech drew attention by tapping his glass, standing up and beginning to speak. This meant, initially at least, that the guests put down their knives and forks and listened, which meant in turn that the hot food got cold and the chilled desserts began to melt. However, as the speeches continued, the guests got tired of waiting and began to eat again.

The order of speeches was changed later to a form that is still adhered to. The laureates are given a fixed number of minutes in which to give their speech of thanks over coffee after the dessert has been served. As the number of guests continued to grow, the event grew out of the Golden Hall and was moved to the Blue Hall below. This hall is not symmetrical, so the staff responsible for laying the tables have to measure the placing of the tables and the covers. If they did not do so, the tables and the covers would appear to go in waves. A special table service called the Nobel service was produced for the 90th anniversary in 1991, and this service was awarded a prize for Excellent Swedish Design the following year. Today, there is hardly any banquet in the world that is so well known as the Nobel Banquet. What is obvious when one reads interviews with the laureates is that nobody has ever complained about the cold weather, the slush, the rain or the icy roads; everyone without exception has nothing but good to say about the event. Two points are mentioned all the time: good conversation and good food.

This book has looked at the menus from December 10 and from the royal palace on December 11 during the 100 years from 1901 to 2001. The court's banquet for the laureates on December 11 is included since it has been held for the laureates ever since 1904. Nobel banquet food has attracted great attention over the years. Right from the start, what is to be served has been kept a close secret; only once during all these years has the menu been published on the same day that the banquet was held, and that was on 10 December, 1945, that is, the first time after the Second World War. This was also the first year that the newspapers wrote about the menu that was to be served at the royal palace on December 11. Previously, there had only been reports that there had been a banquet and who had been there. Until the 1980s, it was the kitchen staff of the Grand Hotel and the City Hall (with some extra staff) who were responsible for composing the menus and preparing the food. Since then, the Chef of the Year Society has been commissioned to plan and prepare the menu in cooperation with some of the permanent staff in the Nobel kitchen at the Stockholm City Hall. However, the mode of operations has always been the same. Around springtime, the work begins to put together a number of menu proposals that are sent to the Nobel Foundation. Later, during the autumn, these menus are test-eaten and the menu is finalised — which then remains a close secret until it is served on the evening of December 10. Until then, only about 30 people know what food is to be offered.

During the first ten years of the banquet, the menus were greatly French-inspired, which is not so surprising considering that the French cuisine has always had a great influence on Swedish cooking. There were up to seven courses on the menu, and throughout the years the banquet started with a glass of dry champagne, which was also served to the dessert. Another constant feature was that sherry was served to the first-course soup, whatever kind of soup it was. This was very tradition-bound because it was not at all unusual at that time to do things that way. A first course that recurred frequently during the first 20 years of the banquets, on both December 10 and December 11, was mock turtle soup. Consommé was also very commonly served and that continued until the 1940s.

It was not until the first banquet after the Second World War, in 1945, that the number of courses was reduced to three: a first course, a main course and a dessert. This has been kept to with only one exception; in 1991, when the 90th anniversary was celebrated, there were four courses. Although the number of courses was reduced, the range of drinks on offer was much more varied and this pattern was maintained until 2001. It was not until the 1980s that people felt that the menu should have a Scandinavian or Nordic profile. The banquet on December 11, on the other hand, has throughout the whole period had more courses than the banquet on December 10. When the present King ascended the throne, a new tradition was started that is still flourishing. Roe deer is not served on December 10 because the court always has it on its menu. The reason is that the King himself is a hunter and fells the roe deer that are served on December 11.

As for the now famous Nobel Glace Parfait, it is interesting to note that in the series of ten-year periods I have used, it was served ten times out of a possible ten only in the period 1981-1990. Chilled desserts like ice-cream and sorbet, however, were common during the second 50-year period. Before that, the desserts consisted of a great deal of fruit, on both December 10 and 11, and pears seem to have been a favourite.

The fact that all the people attending the Nobel banquets over all these years have been successfully served without ever having suffered a cultural clash is something of a *tour de force*. It is an impressive feat considering all the factors that govern gastronomic man.

Not even in his wildest dreams could Alfred Nobel have imagined what he created with his will when he laid the foundations of the Nobel

Prizes and the consequent Nobel Banquet. He could not possibly have foreseen that at the centenary celebrations almost 1,500 guests would be eating a Scandinavian-style menu and taking part in an internationally famous event. All this was achieved by a man who had a French housekeeper in the kitchen and who only liked pancakes with jam from Sweden's own national gastronomy.

With these words, I hand over to someone else the task of surveying and assessing the future Nobel banquets.

Sources and Bibliography

Unpublished sources

Minutes taken at the reading on 14 December 1896 of Alfred Nobel's will.

Palace archives (SLA)

The following are the financial accounts of the royal banquets:
Hovförtäringsräkenskaper, 1904 ID:151.
Hovförtäringsräkenskaper, 1905 ID:153.
Hovförtäringsräkenskaper, 1906 ID:157.
Hovförtäringsräkenskaper, 1908 II.
Hovförtäringsräkenskaper, 1909 II.
Hovförtäringsräkenskaper, 1910 II.
Hovförtäringsräkenskaper, 1911 II.
Hovförtäringsräkenskaper, 1912 II.
Hovförtäringsräkenskaper, 1913 II.
Hovförtäringsräkenskaper, 1920 I-II.
Hovförtäringsräkenskaper, 1921 I-II.
Hovförtäringsräkenskaper, 1922 1-II.
Hovförtäringsräkenskaper, 1923 I-II.
Hovförtäringsräkenskaper, 1925 I-II.
Hovförtäringsräkenskaper, 1926 I-II.
Hovförtäringsräkenskaper, 1927 I-II.
Hovförtäringsräkenskaper, 1928 I-II.
Hovförtäringsräkenskaper, 1929 I-II.
Hovförtäringsräkenskaper, 1930 I-II.
Hovförtäringsräkenskaper 1931 I-II,
These documents and accounts refer to banquets Vols. 24–30.

Royal Household reports

Menu for 11/12 1957.
Menu for 11/12 1960.
Menu for 11/12 1961.
Menu for 11/12 1964 and 1965.
Menu for 11/12 1992.
Menu for 11/12 1994.
Menu for 11/12 1999.

The Nobel Foundation

The Nobel menus for 10 December 1901–2001.

Oral sources

Andersson, Lars-Göran, former restaurateur at Stockholm City Hall.
Einarsson, Lars, head of research for the Kronan project at Kalmar Läns Museum, Sweden.
Lagerfelt, Gunilla, former guest coordinator, The Nobel Foundation, Stockholm.
Löfgren Bo:s sondotter.

The Internet

Nationalencyklopedins website, search words diet, livsmedel kokkonst, kosthåll, gastronomi.
The Nobel Foundation's official website, http://nobelprize.org/.

Magazines

Vin & Sprit Journalen, nr 3-2002, artikel, På gång, Stockholm, 2002.
Vår föda 7/90, Förändrade kostvanor hos invandrare, Kocktürk-Runefors, Tahire, Uppsala, 1990.

Newspaper articles

Numerous articles in the Nobel Foundation's collection of press cuttings from Swedish newspapers.

Unpublished text

Barnekow, Louise & Östberg, Jenny, *Måltiden som upplevelse*, opublicerad C-uppsats i Restaurang och Måltidskunskap, Örebro Universitet, 2000.

Published literature

Algotson, Stina, *Den Svenska maten*, Helsingborg, 1993.
Andersson, Sten, *Matens roller-sociologisk gastronomi*, Stockholm, 1980.
Andrews, Birgitta, *Med smak av Europa*, Stockholm, 1995.
Åstrand, Göran, *Känt och okänt på Stockholms kyrkogårdar*, Stockholm, 1998.
Barett, Christine, *EnBokOm Mexikansk Mat*, Stockholm, 1993.
Beck, Simone, Bertholle, Lousette, & Child, Julia, *Det goda franska köket*, Falun, 1983.
Bettoja, Jo & Corneto, Anna Maria, *Det klassiska Italienska köket*, Helsingborg, 1984.
Brilliant-Savarin, Anthelme, *Smakens fysiologi*, Stockholm, 1958.
Bringéus, Nils-Arvid, *Mat och måltid*, Lund, 1998.
Budd, Jim, *Appreciating Fine Wines*, New Jersey, 1996.
Danforth, Randi, Feierabend, Peter & Chassman, Gary, *Culinaria USA*, Köln, 1999.
Davalos, Julia Elena, *Gran libro de la Cocina Argentina*, Buenos Aires, 1985.
Detrick, Mia, *Sushi-fisk och skaldjur i det japanska köket*, Stockholm, 1988.
Elowson, Hans, *Munskänken*, 1981:2, *Vinet i Gamla testamentet*, Hedemora, 1981.
Ensrud, Barbara, *Ost från hela världen*, Göteborg, 1982.
Flink, Helen, *Orientalisk mat*, Västerås, 1992.
Fréson, Robert, *Det goda Sverige*, Stockholm, 1998.
Grigson, Jane (ed.), *The World Atlas of Food*, London, 1988.
Hafner, Dorinda, *A Taste of Africa*, Hong Kong, 1994.
Halling, Björn, *Måltider och matsedlar*, Värnamo, 1994.
Hartshorn, Jane, *Karibiskt*, Italien, 1997.
Isaksson, Sven, *Vild vikings vivre-om en tidigmedeltida matkultur*, Fornvännen 98 (2003), Stockholm.
Jakobsson, Oskar, *Svenska Landskapsrätter*, Stockholm, 1982.
Johnston, Mireille, *Det provensalska köket*, Stockholm, 1979.
Kocktürk, Tahire, *Kartläggning av matvanor*, Uppsala, 1999.
Kofod, Else Marie, *Skik & Brug*, u.o., 1994.
Kong Foong Ling, *The Food of Asia*, Singapore, 1998.

Kyrlund, Ulli, *Strömming och sill*, Värnamo, 1983.
Lake, Patricia & Barhust Lee, Sharon, *Det mexikanska köket,* Ungern, 1993.
Leeming, Margaret, *Allemans bok om fransk husmanskost*, Stockholm, 1979.
Liukko, Anneli, *Mat, kropp och social identitet*, Stockholm, 1996.
Mendel, Janet, *Det genuina Spanska köket*, Stockholm, 1996.
Mondavi, Robert, *Wines of the World*, London, 1992.
Montaña, Antonio, *El sabor de Colombia*, Bogotá, 1994.
Nordli, Katarina & Harald, *Köksvägen till Portugal*, Stockholm, 1997.
Parker, Jill & Giardino, Patrik, *Den amerikanska kokboken*, Västerås, 1997.
Reece, Eileen, *Det Franska lantköket,* Great Britain, 1987.
Reichert, Wolfgang W., *Germany — A Culinary Tour*, Künzelsau, u.å.
Reynolds, Mary, *Italiensk mat*, u.o., 1981.
Rogov, Daniel, *Dining in the Holy Land 2000 Years Ago*, Jerusalem, 2000.
Rogov, Daniel, *The Diverse Israeli Table*, Jerusalem, 2000.
Schück, Peter, *Fransk husmanskost II*, Stockholm, 1982.
Söderberg, Johan, *Att studera konsumtionen — om långa linjer i tänkandet*, Ekonomin i samhället-ekonomsociologiska perspektiv, Lund, 2002.
Söderlind, Ulrica, *Gastronomiska landskap*, Värdefulla landskap, Umeå, 2003.
Söderlind, Ulrica, *Matvrak- kosthållning och måltidsordning ombord svenska örlogsfartyg under 1600-talet*, licentiatavhandling, Stockholms universitet, 2004.
Stewart, Anita, *Great Canadian Cuisine,* Vancouver, 1999.
Stiernlöf, Annica, *Rysk mat*, Falköping, 1970.
Svenska akademins ordbok, spalt G 149, M449.
Swahn, Jan-Öjvind, *Mathistorisk uppslagsbok*, Värnamo, 1999.
Swahn, Jan-Öjvind, *Fil, fläsk och falukorv — svenska mattraditioner genom tiderna*, Lund, 2000.
Symons, Michael, *A History of Cooks and Cooking*, Devon, 2001.
The Essential Asian Cookbook, Köln, 1998.
Thurfjell, Karsten (text), *Kocklandslagets underbara värld*, Danmark, 2000.
van Wyk, Magdaleen, *Cooking the South African Way,* Capetown, 1986.
Webb, Yvonne, *Det ryska köket*, Australien, 1997.
Westblom-Jonsson, Helen, *Mat, tro, tradition*, Falköping, 1997.
Wilson, Anne, *Irländsk matlagning*, Köln, 1999.
Wolf-Cohen, Elizabeth, *Judisk matlagning — den autentiska kokboken med både moderna och traditionella recept*, Köln, 1998.
Zotterman, Yngve, i Gastronomi, vetenskap, konst, teknik, *Fysiologisk smakforskning, Smakens fysiologi*, Stockholm, 1969.

Appendix 1

The Nobel Banquet Menus (in French), 1901–2001

Food	**Drink**

1901

Food	Drink
Hors d'œuvre Suprême de barbue à la normande Filet de bœuf à l'impériale Gelinottes rôties, salade d'Estrée Succès Grand Hôtel, pâtisserie	Niersteiner 1897 Château Abbé Gorsse 1881 Champagne Crème de Bouzy Doux et Extra Dry Xerez

1902

Food	Drink
Hors d'œuvre Consommé en tasse Homard à la Vanderbildt Tournedos à la mode Perdreaux rôtis, salade d'Alger Succès de glace à la Grand Hôtel Pâtisserie Fruits	Golden Sherry Château Lanessan Johannisberger G. H. Mumm Crémant, Desbordes Brut Porto Sandeman

1903

Food	Drink
Hors d'œuvre Tortue claire en tasse Filets de barbue à l'amiral Chaud-froid de gelinotte à la Lucullus Poulet de Hambourg Salade d'Alger Glace royal Pâtisserie Fruits	Golden Sherry Grand Vin Clos la Garde Hochheimer Duminy, Sec G. H. Mumm Crémant Porto Sandeman

1904

Hors d´œuvre
Consommé en tasse, petits pâtés
Filets de sole à la Duroc
Chaud-froid de poularde à la Jeannette
Faisan rôti
Salade d´Alger
Glace aux fleurs à la Grand Hôtel
Pâtisserie
Fruits

Golden Sherry
Chateau Ripeau
Niersteiner
G. H. Mumm Crémant
Duminy Sec
Porto Sandeman

1905

Hors d´œuvre
Consommé royal
Filet de sole à la gentilhomme
Selle southdown à la duchesse
Chaud-froid de perdreaux à la Lucullus
Salade d´Alger
Fonds d´artichaut à la princesse
Succés de glace à la Marguerite
Pâtisserie
Fruits

Golden Sherry
Château Doutor
Hochheimer
G. H. Mumm Crémant
Romanée
Apollinaris
Porto Sandeman

1906

Hors d´œuvre
Tortue claire en tasse
Suprême de sandre catalane
Poulard à la bonne-femme
Chaud-Froid de cailles à la Lucullus
Salade princesse
Bombe Hélène
Friandises
Fruits

Pale Sherry
Château Grand Vin Clos de la Garde
Johannisberger
G. H. Mumm Crémant
Pol Roger Medium Dry
Porto Sandeman

1907

Banquet cancelled due to the death of King Oscar II.

1908

Hors d´œuvre
Tortue claire à l'anglaise
Filets de sole Marguery
Selle de pré, salé à la choron
Médaillons de homard à la parisienne
Faisan doré rôti
Salade d´Alger
Fonds d´artichaut Maintenon
Succès de glace à la Victoria
  Friandises
Fruits

Madère
Château Palmer
Hochheimer
Roederer, Carte Blanche
Roederer, Grand Vin Sec
Château Lafite
Giesshübler
Porto Sandeman

1909

Consommé Marguérite
Suprême de turbotin Polignac
Filet de bœuf à la renaissance
Poulardes du Mans rôties
Salade duchesse
Haricots verts nouveaux
Poires reine Victoria
  Friandises
Fruits

Pale Sherry
Château Beychevelle
Rüdesheimer
Charles Heidsieck, Sec
Charles Heidsieck, Extra Dry
Château Ranzan Segla
Porto Sandeman

1910

Consommé Alexandra
Filets de sole à l´américain
Selle de pré-salé à la régence
Chaud-froid de gélinotte Lucullus
  Salade niçoise
Asperges vert
  Sauce vierge
Bombe-Nabob
  Friandises
Fruits

Golden Sherry
Pontet Canet
Rüdesheimer
Charles Heidsieck, Impérial
Charles Heidsieck, Extra Dry
Château Lafite
Porto Sandeman

1911

Consommé Doria
Suprême de turbotin cardinal
Poularde fermière
Chaud-froid de cailles
  Salade mâche
Fonds d´artichauts duchesse
Charlotte Râchel
Fruits

Golden Sherry
Chateau Montrose 1896
Liebfraumilch 1904
Charles Heidsieck, Sec, 1900
Charles Heidsieck, Extra Dry, 1900
Porto Sandeman

1912

Consommé zsarine
Suprême de barbue Polignac
Selle d´agneau portugaise
  Sauce béarnaise
Poularde Jeanette
  Salade
Fonds d´artichauts princesse
Parfait Nobel, friandises
Fruits, dessert

Golden Sherry
Château Lafite, 1900
Johannisberger, 1907
Ch. Heidsieck, Demi-sec
Ch. Heidsieck, Brut, 1903
Porto, Very Superior Old

1913

Tortue claire
Suprême de turbotin Walewska
Poularde Massenet
Chaufroix de cailles Lucullus
  Salade
Fonds d´artichauts Maintenon
Parfait praliné
  Friandises
Fruits

Madère Old
Château Smith Haut Lafite, 1905
Rüdesheimer, 1908
Charles Heidsieck, Demi-sec
Charles Heidsieck, Brut, 1904
Porto, Very Superior Old

1914-1919

No banquets were held due to the First World War.

1920 Hasselbacken

Consommé bouquetière
Saumon froid niçoise
Selle de pré-salé renaissance
Asberges en branches
Parfait à l´abricots
Petits fours glacés
Friandises

Ch. Pontet Canet
Forster Traminer, 1893
Sparkling Hock
Fine Bual

1921

Vraie tortue claire
Supremês de turbotin à la Walewska
Selle d´agneau à la renaissance
  Sauce choron
Céleri à la moelle
Poires cardinal
Friandises

Château Paveil de Luze, 1914
Zeltinger Himmelreich, 1913
George Goulet Américain
Madère

1922

Consommé Chesterfield
Piggvarsfilet à la Mornay
Orre à la petit-duc
Kronärtskockbottnar med svampstuvning
Glace-soufflé, småkakor
Frukt, dessert

Madeira Old
Ohligsberger, 1913
S:t Estèphe, 1914
Ayala & Co., Goût Américain
Madeira Old
Portvin Fine Red

1923

A banquet was held but no menu has been preserved.

1924

The banquet was cancelled as no laureates were able to come to Stockholm.

1925

Tortue claire
Suprêmes de sandre Mornay
Dindonneau niçoise
Fonds d´artichauts princesse
Parfait Nelusko
Petits fours
Fruits
Dessert

Madeira Old
Château Tourcas, 1914
Schloss Bärkenheimer Reisling, 1919
V:ve Pommery Américain
Porto

1926

Consommé des gourmets
Filet de barbue maréchal
Selle d´agneau aux légumes
Fonds d´artichauts Maintenon
Ananas à la brésilienne
Petits fours
Fruits
Dessert

Old Madeira
Château Paveil de Luze, 1911
Liebfraumilch, 1922
V:ve George Goulet & Co. Extra Américain
Imperial Ruby

1927

Consommé Chesterfield
Suprême de turbotin normande
Caneton a l´orange
Fonds d´artichauts Mornay
Bombe pralinée
Petits fours
Fruits

Old Madeira
Domaine de Chevalier, 1918
Liebfraumilch Spätlese, 1924
Duminy & Co, Goût Américain
Imperial Ruby

1928

Consommé trois filets
Turbot à l´ostendaise
Gelinotte à la Lucullus
Fonds d´artichauts à la Riche
Soufflé glacé royal
Petits fours
Fruits
Dessert

Old Madeira
Ch. L´Evangile, 1920
Liebfraumilch, Spätlese, 1924
Duminy & Co, Goût Américain
Imperial Ruby

1929

Consommé Westmoreland
Filets de sole au champagne
Caneton froid à la voisin
Salade de saison
Fonds d´artichauts à la Royal
Bombe glacée aux noisettes
Petits fours

Madeira Fine
Clos Fourtet, 1922
Niersteiner Weissenberg, 1926
J. Bollinger, Dry Extra
Imperial Ruby

1930

Tortue claire en tasse
Truits au bleu à la financière
Dindonneau rôti, salade et gelé
Fonds d´artichauts à la Paiva
Parfait à la Tosca
Petits fours
Fruits assortis
Café

Madeira (Old Stock)
Niersteiner Rote Schmitt, 1926
Champagne (G. H. Mumm et Cie)
Extra Portwine (Vintage)

1931

Crème châtelaine
Filets de sole Joinville
Caneton braisé aux céleris
Soufflé glacé au Curaçao
Petits fours
Fruits

Golden Ordinary
Niersteiner Weissenberg, 1928
Ch. Beycheville, 1924
Louis Roederer, Grand Vin Sec

1932

Tortue claire
Turbot à l´amiral
Caneton à la royal
Soufflé glacé aux violettes
Petits fours
Fruits

Golden Sherry
Schloss Böckelheimer, 1928
Ch. Beychevelle, 1924
Louis Roederer, Grand Vin Sec

1933

Consommé des gourmets
Filets de sole maire
Poularde de Sandemar poêlée à la renaissance
Bombe glacée Nesselrode
Corbeilles de petits fours
Fruits
Friandises

Golden Sherry
Schloss Böckelheimer, 1929
Ch. Lagrange, 1924
Loius Roederer, Grand Vin Sec

1934

Tortue claire
Filets de sole à la Germaine
Faisan froid à la strasbourgeois
Bombe glacée à la Rotschild
Courbeilles de petits fours
Fruits
Friandises

Amontillado Superior
Kreuznacher Spätlese, 1929
Château Palmér, 1924
Louis Roederer, Grand Vin Sec
Pommery & Greno, Drapeau Vin Sec

1935

Consommé Chesterfield
Filet de sole special
Selle de chevreuil, Lucullus
Parfait au mocca aux petits fours
Fruits
Friandises

Amontillado N. P. U.
Château Pontet Canet, 1924
Louis Roederer, Grand Vin Sec
Pol Roger, Brut, 1926
Fred´s Madeira, Old Stock

1936

Consommé de tortue claire
Filets de sole
Suprême de gelinotte à la Florida
Salade Waldorf
Glace au four
Fruits assortis
Friandises

Amontillado Superior
Pol Roger, Brut Réserve
Louis Roederer, Grand Vin Sec
Château Montrose, 1924
Fred´s Port, Dark Stork

1937

Consommé Chesterfield
Filet de barbue à la picarde
Caneton froid à la châtelaine
Salade russe
Parfait glacé au nougat
Petits fours
Fruits assortis
Friandises

Fred´s Sherry
Pol Roger sec
Louis Roederer demi-sec
Grand Vin le Valon-Hanappier, 1929
Vintage Port, 1924

1938

Consommé trois filets	Sherry, El Caballero
Filet de sole, froid à la norvégienne	Pol Roger, Brut Réserve
Poularde rôtie à la châtelaine	Louis Roederer, Gran Vin Sec
Salade de saison	Grand Vin le Vallon-Hanappier 1929
Soufflé glace pralinée	Fred´s Port
Petits fours	
Fruits assortis	
Friandises	

1939-1944

No banquets were held due to the Second World War.

1945

Potage aux champignons	Sherry Granja
Selle de renne suédoise	Douro Clarete
Haricots-verts au beurre battu	Imperial Ruby
Bombe-glacée Stadshuset	
Friandises	

1946

Tortue claire en tasse	Sherry Granja
Selle de renne suédoise	Mendoza Tinto
Pommes parisienne	Imperial Ruby
Fonds de céleri	
Poire glace Stadshuset	
Friandises	

1947

Sandwiches	Vin Rouge
Poulet fermière	Sherry
Gateau de pommes	
Sauce vanilla	

1948

Consommé Chesterfield	Amontillado
Helstekt oxfilé Polignac	Margaux
Glace päron Belle Hélène	

1949

Crème d´Argenteuil
Oison à la suédois aux choux-rouge
crème de raifort
Gateaux Napoleon

Amontillado Superior
Marquisat

1950

Potage aux champignons
Filet de bœuf béarnaise
Fonds d´artichauts princesse
Glace Nobel
Friandises

Kressman Monopole Rouge
Champagne Perrier-Jouët & Cie
Café Avec

1951

Homard parisienne
Filet de veau Polignac
Pommes risollées
Glace Nobel
Friandises

Graves Rosechâtel
King's Port

1952

Pirogue au saumon
Beurre noisette
Coq de bruyère rôti
Salade
Pommes parisienne
Fruit en espalier

Kressmann Monopole Rouge
Magnum
Old Superior White Port

1953

Homard à la parisienne
Pain riche
Selle de chevreuil rôtie
Sauce grand veneur,
purée de marrons
Savarin aux fruits à la liqueur de
Maraschino

Deutz Champagne Brut
Cachet Nobel
Château Kirwan en Magnum
Eau minérale Ramlösa
Café
Cognac Frapin V.S.O.P.
Liqueurs

BUFFET

1954

Truite de rivière fumée
  Epinards à la crème
Filet de bœuf rôti
  Fonds d´artichauts princesse
  Champignons frais au Porto
  Pommes parisienne
Poires Roberta
	Ockfener Bocksteiner, 1952
	Château Gruaud Larose 1950,
	  en Magnum
	Heidsieck Brut Dry Monopole
	Eau minérale Ramlösa
	Café
	Hine V.O.
	Benedictine
	
	BUFFET

1955

Bisque de homard
Filet de bœuf rôti à la Bordelaise
  Pommes parisienne
Bavarois aux fruits, sucre filé
  Friandises
	Amontillado Pale
	Château la Tour de Mons, 1952
	Champagne Eugène Barbier
	Café
	Cognac Rémy Martin V.S.O.P.
	Cointreau Liqueur
	
	BUFFET

1956

Saumon au vin rouge à la Madrid
Dinde en gelée, sauce au vin
  Salade mimosa
Parfait aux mûres de Laponie
  Feuilles d´amandes
	Niersteiner Gutes Domthal Spätlese, 1949
	Château d´Angelus, 1953
	Lanson Père & Fils Brut, Carte Noire
	Café
	Cognac Rémy Martin V.S.O.P.
	Cointreau Liqueur
	
	BUFFET

1957

Truite de rivière à la parisienne
  Sauce verte
Faisan au Porto
  Laitue romaine
Poires au Cognac Fine Champagne
  Crème frappée
	Johannisberger Riesling, 1955
	Château Lescadres
	Deinhardt Cabinet, Brut
	Café
	Cognac Rémy Martin V.S.O.P.
	Cointreau Liqueur
	
	BUFFET

1958

Filet de sole bourguignonne
Chaud-froid de caneton au Porto
  Salade mimosa
  Sauce orange
Parfait aux avelines
	Château Liversan, Medoc 1953
	Lanson Dry Extra
	Café
	Cognac Martell
	Liqueur Bols Apricot Brandy
	
	BUFFET

1959

Suprême de turbot à la fine champagne
Riz au beurre
Poularde froide à la gelée
Salade Waldorf
Poires flambées à la frangipane

Château Lescadre, Medoc, 1955
Champagne Moët et Chandon, Brut S.A.
Café
Liqueur Benedictine D.O.M.
Armagnac V.S.O.P

BUFFET

1960

Potage trois couronnes
Selle d´agneau rôtie, sauce au madère, pimentée
Pommes rissolées
Salade de saison
Savarin lapon au rhum
Crème fouettée

Château de Plaisan, 1955
Champagne Pommery & Greno Brut
Café
Cognac Frapin V.S.O.P.
Liqueur Benedictine D.O.M.

BUFFET

1961

Filets de sole au vin blanc
Riz au curry
Coq des bois et coq de bruyère froids en gelée
Sauce périgourdine
Salade Waldorf
Poires à la fine champagne
Sauce à la vanille

Pommery & Greno Brut
Château Chauvin, 1955, S:t Emilion
Café
Liqueur Benedictine D.O.M.
Cognac Frapin V.S.O.P.

BUFFET

1962

Truite de rivière fumée à la parisienne
Poularde rôtie, sauce madère au foie gras
Pommes rissolées
Salade de saison
Pêche au Grand Marnier
Crème Chantilly

Château Bellevue, 1955, S:t Emilion
Pommery & Greno Brut
Café
Liqueur Marie Brizard & Strega
Cognac Courvoisier

BUFFET

1963

Saumon braisé à la madrilène
Caneton froid à la gelée
Sauce périgourdine
Salade de mandarines
Poires Nobel Gala

Château Bellevue, 1955, S:t Emilion
Pommery & Greno Brut
Café
Liqueurs Marie Brizard & Strega
Cognac Courvoisier

BUFFET

1964

Turbot fumé à chaud
 Sauce gourmet
Chaud-froid de gelinotte au foie gras
 Salade Waldorf
Pêches au Grand Marnier
 Petits fours

Pommery & Greno Brut
Château des Tonelles 1959, Côtes Fronsac
Café
Liqueur Marie Brizard
Cognac Courvoisier

BUFFET

1965

Paupiettes de sole pochées à la Trois
 Couronnes
Chaud-froid de poularde farcie
 Asperges vinaigrette
 Sauce madère au foie gras
Ananas à la macédoine aux liqueurs
 Petits fours

Pommery & Greno Brut
Château du Basque 1959
Café
Liqueur Marie Brizard
Cognac Courvoisier

BUFFET

1966

Turbot fumé à chaud
 Sauce gourmet
Perdreau à la bohémienne
 Salade mimosa
Glace à la vanille
 Sauce chocolat
 Friandises

Pommery & Greno Brut
Château les Chaumes 1960,
 Côtes de Blaye
Café
Liqueurs Marie Brizard & Strega
Cognac Courvoisier

BUFFET

1967

Saumon fumé
 Fond d´artichauts aux épinards
 Crème raifort
Canard sauvage provençale
 Sauce madère et salade
Poire Nobel Gala

Pommery & Greno Brut
Château Laurensanne 1961
Café
Liqueurs Marie Brizard & Strega
Cognac Courvoisier

BUFFET

1968

Avocats au homard
 Sauce gourmet
Selle d´agneau, morilles à la crème
 Sauce madère et salade Waldorf
Parfait glace à l´ananas
 Petits fours

Pommery & Greno Brut
Château d´Eyquem 1962
Café
Liqueurs Crème de Cacao Brizard & Strega
Cognac Courvoisier

BUFFET

1969

Avocats farcis au saumon
  Sauce gourmet moscovite
Filet de bœuf rôti périgourdine
Sorbet à l´orange

Krug Brut Réserve
Château Potensac 1959
Café
Cognac Courvoisier
Liqueurs Crème de Cacao Brizard & Strega

BUFFET

1970

Saumon fumé à chaud
  Fond d´artichaut, epinards
  Sauce mousseline
Selle de renne farcie
  Sauce aux genièvres
  Pommes dauphines
  Salade
Savarin aux fausses-mûres
  avec crème moka

Krug Brut Réserve
Château Mayne-Vieil 1959
Café
Cognac Renault Carte Noire
Liqueur Crème de Cacao Brizard & Strega

BUFFET

1971

Mousse froide de turbot truffée
  au champagne à l´Edlund
Poulet rôti Christineberg
  Salade
Soufflé glace Grand-Marnier à la Solliden

Krug Brut Réserve
Château Grand Barrail Lamarzelle
  Figeaz 1962
Café
Cognac Renault Carte Noir
Liqueur Apricot Brandy Bols
Cherry Heering

BUFFET

1972

Turbot fumé à chaud
  Sauce hollandaise
Gigot d´agneau rôti, sauce madère
  Salade tomates
Glace à la suédoise

Krug Brut Réserve
Château Plaisance 1966
Café
Cognac Renault Carte Noire
Liqueur Benedictine

BUFFET

1973

Saumon fumé
  Œufs pochés
Filet de bœuf Charlemagne
Parfait à l´ananas

Krug Private Cuvée
Château Laurensanne 1966
Café
Cognac Renault Carte Noire Extra
Liqueur Peter Heering

BUFFET

1974

Saumon fumé à chaud
  Sauce hollandaise
Cuissot de chevreuil rôti
  Sauce genièvre
  Salade
Sorbet à l´orange

Krug Private Cuvée
Château Landreau 1970
Café
Cognac Renault Carte Noire Extra
Liqueur Benedictine

BUFFET

1975

Mousse froide de barbue
  Sauce aux écrevisses
Poule de neige rôtie
  Sauce au foie gras et truffes
  Gelée de sorbes
  Pommes Trois Couronnes
  Salade de saison
Parfait glace aux airelles rouges
  Biscuits aux amandes

Krug Brut Private Cuvée
Château Lacaussade 1970
Café
Cognac Renault Carte Noire Extra
Liqueur Mandarine Napoléon

BUFFET
Gordon´s Gin
Bell´s Whisky

1976

Filets de sole pochés au vermouth
  Riz pilaf
Mousse froide de gelinotte
  Sauce périgueux
  Salade Waldorf
Parfait Glace Nobel
  Petits fours

Pommery & Greno, Brut
Château Lacaussade 1970
Café
Cognac Bisquit Dubouche ***
Liqueur Mandarine Napoléon

BUFFET
Gordon´s Gin
Black and White Whisky

1977

Pâté de saumon
  Sauce verte
Poule de neige braisée aux genièvres
  Gelée de sorbes
  Salade de saison
Parfait Glace Nobel
  Petits fours

Pommery & Greno, Brut
Château Trinité-Valrose, 1970
Café
Cognac Martell ***
Liqueur Mandarine Napoléon

BUFFET
Gordon´s Gin
Stewart, Finest Old Scotch

1978

Homard parisienne
Caneton braisé au Calvados
  Compote de pommes
Parfait Glace Nobel
  Petits fours

Pommery & Greno, Brut
Château Trinité-Valrose 1970
Café
Cognac Bisquit Dubouche***
Liqueur Mandarine Napoléon

BUFFET
Bols Silver Top Dry Gin
Bell's Scotch Whisky

1979

Suprêmes de turbot froid aux œufs
  d'ablette
  Sauce hollandaise
Selle de veau rôtie
  Sauce morilles
  Pommes sautées
  Salade de haricots verts
Parfait Glace Nobel
  Petits fours

G. H. Mumm, Cordon Rouge, Brut
Château Trinité-Valrose, 1970
Eau minérale Ramlösa
Café
Baron Otard, Fine Champagne, V.S.O.P.
Liqueur Mandarine Napoléon

BUFFET
Bols Silver Top Dry Gin
Long John Whisky

1980

Saumon fumé aux épinards
  Œuf poché
Filet de renne aux chanterelles
  Sauce Akvavit, pommes lyonnaise
  Salade et gelée
Parfait Glace Nobel
  Petits fours

G. H. Mumm Cordon Rouge, Brut
Château Landreau 1976
Eau minérale Ramlösa
Café

BUFFET
Long John Whisky
Bols Silver Top Dry Gin
Campari

1981

Mousse de saumon
  Sauce d'écrevisses ou sauce verte
Filet d'élan aux choux de Bruxelles
  Sauce aux morilles, pommes poireau
  Confiture de sorbe
Parfait Glace Nobel
  Petits fours

G. H. Mumm, Cordon Rouge, Brut
Château Saint Germain, 1975
Eau minérale Ramlösa
Café

BUFFET
Long John Whisky
Bols Silver Top Dry Gin
Campari, Cachucha, Nutty Solera

1982

Filet de renne mariné
Sauce Dijon
Omble-chevalier braisé à la crème d'aneth
Riz
Parfait Glace Nobel
Petits fours

G. H. Mumm Cordon Rouge, Brut
Kopke Bridge
Eau minérale Ramlösa
Café

BUFFET
Long John Whisky
Bols Silver Top Dry Gin
Campari, Cachucha, Tio Pepe

1983

Mousse de perdrix de neige
Sauce à la crème
Filet de sole farci aux morilles
Sauce Noilly, riz
Parfait Glace Nobel
Petits fours

Château Montfaucon, 1979
Pommery & Greno
Eau minérale Ramlösa
Café

BUFFET
Bell's Old Scotch Whisky
Bols Silver Top Dry Gin
Campari
Boissons de Spendrup

1984

Saumon mariné aux genièvres
Pâté aux morilles, sauce moutarde
Canard sauvage rôti, sauce aux cassis
Gâteau aux pommes de terre et carottes
Parfait Glace Nobel
Petits fours

Moët et Chandon, Brut Impérial
Château Vieille Tour, 1979
Eau minérale Ramlösa
Café

BUFFET
The Famous Grouse
Queen Anne
Gordon's Dry Gin
Carpano Punt e Mes
Boissons de Spendrup

1985

Timbale d'écrevisses, Sauce basilic
Petit pain Nobel
Selle d'agneau rôtie garnie de fenouil
Haricots verts et toFoodes grillées
Sauce cognac à la crème,
Pommes allumettes
Parfait Glace Nobel
Petits fours

Moët et Chandon, Brut Impérial
Médoc Baron Philippe 1979, Sélection Nobel
Eau minérale Ramlösa
Café

BUFFET
The Famous Grouse
Queen Anne
Gordon's Dry Gin
Carpano Punt e Mes
Boissons de Spendrup

1986

Saumon fumé de Halmstad
Epinards et sauce au caviar rouge du pays
Petit pain Nobel
Faisan rôti Julienne, surprise du bois
Sauce à l'anglaise, pommes parisienne
Parfait Glace Nobel
Petit fours

Moët et Chandon, Brut Impérial
Médoc Baron Philippe 1982, Sélection Nobel
Cockburn's 1981, Late Bottled Vintage
Eau minérale Ramlösa
Café

BUFFET
The Famous Grouse, Whisky
Queen Anne, Whisky
Gordon's Dry Gin
Boissons de Spendrup

1987

Pâté de homard truffé
Sauce au cresson alénois
Petit pain Nobel
Selle de lièvre rôtie, pommes farcies de légumes,
Sauce Calvados et salade verte
Parfait Glace Nobel
Petits fours

Moët et Chandon Brut Impérial
1982 Medoc, Baron Philippe de Rothschild, Sélection Nobel
1981 Cockburn's, Late Bottled Vintage
Eau minérale Ramlösa
Café

BUFFET
The Famouse Grouse Whisky
Glenlivet Malt Whisky 12 Years Old
Gordon's Dry Gin
Hennessy V.S.
Boissons de Pripps

1988

Omble-chevalier mariné
Sauce à la purée de pois verts
Petit pain Nobel
Grand cerf farci à la nordique,
sauce aux morilles
et chanterelles,
légumes de la saison
Parfait Glace Nobel
Petits fours

Moët & Chandon Brut Impérial
1981, St. Estephe, Baron Philippe de Rothschild, Sélection Nobel
Vintage Character Port
Eau minérale Ramlösa
Café

BUFFET
Petite Liqueur par Moët & Chandon
Johnnie Walker Red & Black Label
Gordon's Dry Gin, Bacardi Rum
Hennessy V.S.
Boissons de Pripps

1989

Pâté d´anguille fumée et de sole
Sauce aux œufs d´ablette
Petit pain Nobel
Dos d´élan mariné
Sauce d´airelles à la crème
Julienne de légumes à la bretonne
Parfait Glace Nobel
Petits fours

Moët & Chandon Brut Impérial
Medoc, Baron Philippe de Rothschild 1982, Sélection Nobel
Eau minérale Ramlösa
Café

BUFFET
Petite Liqueur par Moët & Chandon
Johnnie Walker Red & Black Label
Ron Bacardi
Absolut Vodka
Hennessy V.S.
Boissons de Pripps

1990

Timbale de perdix de neige
sur lit de salade à la Raineri
Sauce aux herbes, croissant Nobel
Omble-chevalier suédois
grillé à la genièvre avec asperge beurrée
Crème à l´aneth et riz
Parfait Glace Nobel
Petits fours

Moët & Chandon Brut Impérial
Château de Cruzeau 1988
Grådask Very Old Superior
Vin sans alcool – St Regis
Eau minérale Ramlösa
Café

BUFFET
Absolut Vodka
Petite Liqueur par Moët & Chandon
Johnnie Walker Red & Black Label
Ron Bacardi
Gordon´s Gin
Hennessy V.S.
Boissons de Pripps

1991

Potage aux orties
Croissant Nobel
Tartare du saumon mariné, crème au paprika rouge
Poitrine de canard rôti, sauce aigrette aux baies d´archipel
Compote de plantes-racines à la Raineri
Glace Nobel à la vanille et aux myrtilles
Petits fours

Moët & Chandon Brut Impérial
Tokay-Pinot Gris 1990
Château Mouton Baronne Philippe 1984
Cru Classé – Sélection Nobel
Château Malagar 1990
Premières Côtes de Bordeaux, Maison Cordier
Vin sans alcool – St Regis
Eau minérale Loka
Café

BUFFET
Absolut Vodka
Petite Liqueur par Moët & Chandon
Johnnie Walker Red & Black Label
Ron Bacardi
Gordon´s Gin
Hennessy V.S.
Boissons de Spendrup

1992

Terrine de saumon et de sole à l´aneth
Sauce aux œufs d´ablette
Faux filet d´agneau rôti, champignons de saison
Carottes et salsifis noirs glacés au miel
Sauce au cidre
Glace Nobel au chocolat blanc
et à la groseille rouge

Moët & Chandon Brut Impérial
Château Mouton Baronne Philippe 1984
Cru Classé — Sélection Nobel
Johannisberg Botrytis Riesling 1983
Eau minérale Ramlösa
Café
BUFFET
Absolut Vodka
Hennessy V.S.O.P. Privilège
Petite Liqueur par Moët & Chandon
Johnnie Walker Red & Black Label
Bacardi Rum
Gordon´s Gin
Campari
Boissons de Pripps

1993

Omble mariné sur lit de radis aux œufs de saumon
Vinaigrette de miel à la Raineri
Croissant Nobel
Faux-filet de renne aux chanterelles des forêts
Terrine de choux au beurre de mûres fraîches
Mousse de pommes de terre et emincé de salsifis
Glace Nobel en voile
Sorbet aux fraises des bois et son parfait vanille

Moët & Chandon Brut Impérial 1988
1988 Château Tour St-Christophe
Saint Emilion Grand Cru
Sauternes Baron Philippe de Rothschild 1991
Eau minérale Ramlösa
Café
BUFFET
Absolut Vodka
Hennessy V.S.O.P. Privilège
Petite Liqueur par Moët & Chandon
Johnnie Walker Red & Black Label
Bacardi Rum
Gordon´s Gin
Campari
Boissons de Pripps

1994

Roulade de poitrine de canard fumée à la mangue
et à la bette
Accompagnée de mâche aux pignons
et de son croissant Nobel
Filet de veau parfumé à la sauge
et ses champignons Pom Pom
Tomate farcie et ses épinards garniture
et crêpes de pommes de terre au gingembre
Parfait Nobel en voile
Sorbet aux framboises et son parfait vanille

Moët & Chandon, 1983
Cuvée Spéciale 250ème Anniversaire
Château Liversan, 1990
Haut-Médoc
Moulin Touchais, 1984
Côteaux du Layon
Eau minérale Carl von Linné
Café
DIGESTIFS
Hennessy V.S.O.P. Privilège
Calvados Boulard Hors d´Age
Johnnie Walker Black Label
Gordon´s Dry Gin
Campari
Bacardi Premium Black
Absolut Vodka

1995

Aiglefin de la Mer du Nord delicatement mariné
sur un lit de cresson accompagné de pommes en salade
Croissant Nobel
Filet de cerf fondu dans une sauce de cassis
Garniture de cèpes, courges d´Italie et noix de porc fumé
Petites galettes de pommes de terre
Parfait Nobel en voile
Sorbet aux baies d´archipel et son parfait vanille

1989 Champagne Taittinger
Brut Millésimé
Domaine de la Gaffelière 1982
St Emilion 1976
Lenz Moser Trockenbeerenauslese
Eau minérale Ramlösa
Café

DIGESTIFS
Hennessy V.S.O.P.
Calvados Boulard
Drambuie
Johnnie Walker Black Label
Lagavulin 16 Years Old Single Malt
Campari Bitter
Finlandia Vodka
Gordon´s Gin
Boissons de Pripps

1996

Homard en gelée à la crème de choux-fleur
et au caviar de Kalix
Petit pain Nobel aux 4 céréales
Ballotine de pintade avec sa garniture
de pommes de terre de Laponie et de plantes du potager
Sauce au citron
Glace Nobel
Panache de sorbet aux mûres sauvages de champs
et de parfait à la vanille
Petits fours

Champagne Pommery 1990
Brut Millésimé
Chambolle-Musigny 1:er cru, 1992
Joseph Drouhin
Château Lafaurie-Peyraguey 1990
Domaine Cordier
Eau minérale Ramlösa
Café

DIGESTIFS
Absolut Vodka
Hennessy V.S.O.P.
Calvados Boulard
Drambuie
Johnnie Walker Black Label
Jack Daniels
Gordon´s Gin
Campari Bitter
Boissons de Pripps

1997

Galette aux topinambours de saumon fumé
de Landskrona et de homard
Poitrine de pigeonneau sauvage et petit ragoût
aux bolets à la compote de pommes de terre
et d´oignons, sauce aigre-douce au v
inaigre de framboise
Glace Nobel
Parfait glace à la fleur de sureau et au
sorbet à la fraise

Champagne Pommery 1991, Brut Millésimé
Crozes-Hermitage Domaine de Thalabert 1990
Paul Jaboulet Aîné
Château Pajzos 1993
Tokaji Aszu 5 Puttonyos
Eau minérale Ramlösa
Café

DIGESTIFS
Absolut Vodka
Hennessy V.S.O.P. Privilège
Calvados Boulard
Drambuie
Campari Bitter
Gordon´s Gin
Bell´s 8 years old
Johnnie Walker Black Label
Jack Daniels
Boissons de Pripps

1998

Fonds d´artichaut marinés et garnis aux crevettes,
aux écrevisses et au fenouil
Suprême de poulet au thym et au
poivre noir de Java
Ragoût de légumes de saison et petit rouleau aux
champignons, Sauce à la crème de topinambours
Glace Nobel
Glace au chocolat blanc
et sorbet aux mûres sauvages

Champagne Pommery 1991
Brut Millésimé
Corton Les Bressandes Grand Cru 1989
Prince Florent de Mérode Ladoix
Sandeman´s Tawny Port aged 30 years
Café
Grönstedts Extra et Cederlunds Caloric
Eau minérale Ramlösa

1999

Crème de topinambours au homard et caviar
du pays
Filet d´agneau en robe de choux au
jus aigre-doux à l´aneth,
purée fine de pommes de terre au cheddar
et petits légumes
Glace Nobel
Glace d´épices et sorbet à l´ananas
avec compote de fruits de la passion
et d´ananas

Champagne Pommery 1992
Brut Millésimé
Château de Beaucastel 1996
Chateauneuf-du-Pape
Rivesaltes Amber 1974 Hors d´Age
Café
Grönsteds Extra et Cederlunds Caloric
Eau minérale Ramlösa

2000

Terrine de flétan et langoustines,
coquilles Saint Jacques marinées
et crème fraîche aux œufs d´ablettes
Magret de canard rôti au citron,
timbale de canard et de fromage, ragoût
des girolles et topinambours et purée de
pommes de terre aux herbes
Glace Nobel
Glace à la vanille et sorbet d´airelles
Petits gâteaux aux amandes

Champagne Pommery 1992
Brut Millésimé
Château Larrivet-Haut-Brion 1996
Pessac-Leognan
Château Raymond-Lafon 1987
Sauternes
Café
Grönsteds XO
Facile Punsch
Eau minérale Ramlösa

2001

Homard sur purée de chou-fleur
et gelée de langoustines, salade de salicornes
Caille farcie au foie gras et sa poêle des cèpes
et tomates sechées, asperges vertes et purée
de cerfeuil
Duo des glace vanille et parfait cassis meringue

Louise Pommery Champagne 1989
Château Palmer Margaux 1997
Bernkasteler Badstube Riesling Eiswien 1998
Mosel-Saar-Ruwer
Café
Grönstedts Rarissime
Grand Champagne
Cointreau
Eau minérale Ramlösa

Appendix 2

Nobel Laureates, 1901–2001

Physics

1901 Wilhelm Röntgen, Germany
1902 Hendrik A. Lorentz, The Netherlands; Pieter Zeeman, The Netherlands
1903 Henri Becquerel, France; Pierre Curie, France; Marie Curie, France
1904 Lord Rayleigh, UK
1905 Phillip Lenard, Germany
1906 J. J Thomson, UK
1907 Albert A. Michelson, USA
1908 Gabriel Lippmann, France
1909 Guglielmo Marconi, Italy; Ferdinad Braun, France
1910 Johannes Diderik van der Waals, The Netherlands
1911 Wilhelm Wien, Germany
1912 Gustaf Dalén, Sweden
1913 Heike Kamerlingh-Onnes, The Netherlands
1914 Max von Laue, Germany
1915 William Bragg, UK; Lawence Bragg, UK
1916 No award
1917 Charles Glover Barkla, UK
1918 Max Planck, Germany
1919 Johannes Stark, Germany
1920 Charles Edouard Guillaume, Switzerland
1921 Albert Einstein, Germany and Switzerland
1922 Niels Bohr, Denmark
1923 Robert A. Millikan, USA
1924 Manne Seigbahn, Sweden
1925 James Franck, Germany; Gustav Hertz, Germany
1926 Jean Baptiste Perrin, France
1927 Arthur H. Compton, USA; C.T.R. Wilson, UK
1928 Owen Willans Richardson, UK
1929 Louis de Broglie, France
1930 Venkata Raman, India
1931 No award
1932 Werner Heisenberg, Germany
1933 Erwin Schrödinger, Austria; Paul A.M. Dirac, UK
1934 No award
1935 James Chadwick, UK
1936 Victor F. Hess, Austria; Carl D. Anderson, USA
1937 Clinton Davisson, USA; George Paget Thomson, UK
1938 Enrico Fermi, Italy
1939 Ernest Lawrence, USA
1940 No award
1941 No award
1942 No award
1943 Otto Stern, USA

1944 Isidor Isaac Rabi, USA
1945 Wolfgang Pauli, Austria
1946 Percy W. Bridgman, USA
1947 Edward V. Appelton, UK
1948 Patrick M. S. Blackett, UK
1949 Hideki Yukawa, Japan
1950 Cecil Powell, UK
1951 John Cockcroft, UK; Ernest T. S. Walton, Ireland
1952 Felix Bland, USA; E.M. Purcell, USA
1953 Frits Zernike, The Netherlands
1954 Max Born, UK; Walther Bothe, Germany
1955 Willis E. Lamb, USA; Polykarp Kusch, USA
1956 William B. Shockley, USA; John Bardeen, USA; Walter H. Brattain, USA
1957 Chen Ning Yang, China; Tsung-Dao Lee, China
1958 Pavel A. Cherenkov, Soviet Union; Ilya M. Frank, Soviet Union; Igor Y. Tamm, Soviet Union
1959 Emilio Segrè, USA; Owen Chamberlain, USA
1960 Donald A. Glaser, USA
1961 Robert Hofstadter, USA; Rudolf Mössbauer, Germany
1962 Lev Landau, Soviet Union
1963 Eugene Wigner, USA; Maria Goeppert-Mayer, USA; J. Hans D. Jensen, Germany
1964 Charles H. Townes, USA; Nicolay G. Basov, Soviet Union; Alexsandr M. Prokhorov, Soviet Union
1965 Sin-Itiro Tomonaga, Japan; Julian Schwinger, USA; Richard P. Feynman, USA
1966 Alfred Kastler, France
1967 Hans Bethe, USA
1968 Luis Alvares, USA
1969 Murray Gell-Mann, USA
1970 Hannes Alfvén, Sweden; Louis Néel, France
1971 Dennis Gabor, UK
1972 John Bardeen, USA; Leon N. Cooper, USA; Robert Schrieffer, USA
1973 Leo Esaki, Japan; Ivar Giaever, USA; Brian D. Josephson, UK
1974 Martin Ryle, UK; Antony Hewish, UK
1975 Aage N. Bohr, Denmark; Ben R. Mottelson, Denmark; James Rainwater, USA
1976 Burton Richter, USA; Samuel C. C. Ting, USA
1977 Philip W. Anderson, USA; Nevill F. Mott, UK; John H. Van Vleck, USA
1978 Piotr Kapitsa, Soviet Union; Arno Penzias, USA; Robert Woodrow Wilson, USA
1979 Sheldon Glashow, USA; Abdus Salam, Pakistan; Steven Weinberg, USA
1980 James Cronin, USA; Val Fitch, USA
1981 Nicolaas Bloembergern, USA; Arthur L. Schawlow, USA; Kai M. Siegbahn, Sweden
1982 Kenneth G. Wilson, USA;
1983 Subramanyan Chandrasekhar, USA; William A. Fowler, USA
1984 Carlo Rubbia, Italy; Simon van der Meer, The Netherlands
1985 Klaus von Klitzing, East Germany
1986 Ernst Ruska, Germany; Gerd Binning, Germany; Heinrich Rohrer, Switzerland
1987 J. George Bednorz, Germany; K. Alex Müller, Switzerland
1988 Leon M. Lederman, USA; Melvin Schwartz, USA; Jack Steinberger, USA
1989 Norman F. Ramsey, USA; Hans G. Dehmelt, USA; Wolfgang Paul, Germany
1990 Jerome I. Friedman, USA; Henry W. Kendall, USA; Richard E. Taylor, Canada
1991 Pierre-Gilles de Gennes, France

1992 Georges Charpak, France
1993 Russell A. Hulse, USA; Joseph H. Taylor Jr, USA
1994 Bertram N. Brockhouse, Canada; Clifford G. Shull, USA
1995 Martin L. Perl, USA; Frederick Reines, USA
1996 David M. Lee, USA; Douglas D. Osheroff, USA; Robert C. Richardson, USA
1997 Steven Chu, USA; Claude Cohen-Tannoudji, France; William D. Phillips, USA
1998 Robert B. Laughlin, USA; Horst L. Störmer, Germany; Daniel C. Tsui, USA
1999 Gerardus ´t Hooft, The Netherlands; Martinius J. G. Veltman, The Netherlands
2000 Zhores I. Alferov, Russia; Herbert Kroemer, Austria; Jack S. Kilby, USA
2001 Eric A. Cornell, USA; Wolfgang Ketterle, Austria; Carl E. Wieman, USA

Chemistry

1901 Jacobus Henricus van' t Hoff, The Netherlands
1902 Hermann Emil Fischer, Germany
1903 Svante August Arrhenius, Sweden
1904 Sir William Ramsay, UK
1905 Johann Friedrich Wilhelm Adolf von Baeyer, Germany
1906 Henri Moissan, France
1907 Eduard Buchner, Germany
1908 Ernest Rutherford, UK
1909 Wilhelm Ostwald, Germany
1910 Otto Wallach, Germany
1911 Marie Curie, France
1912 Victor Grignard, France; Paul Sabatier, France
1913 Alfred Werner, Switzerland
1914 Theodore William Richards, USA
1915 Richard Martin Willstätter, Germany
1916 No award
1917 No award
1918 Fritz Haber, Germany
1919 No award
1920 Walther Hermann Nernst, Germany
1921 Fredrick Soddy, UK
1922 Francis William Aston, UK
1923 Fritz Pregl, Austria
1924 No award
1925 Richard Adolf Zsigmondy, Germany
1926 Theodor Svedberg, Sweden
1927 Heinrich Otto Wienland, Germany
1928 Adolf Otto Reinhold Windaus, Germany
1929 Arthur Harden, UK; Hans Karl August Simon von Euler-Chelpin, Sweden
1930 Hans Fischer, Germany
1931 Carl Bosch, Germany; Freidrich Bergius, Germany
1932 Irving Langmuir, USA
1933 No award
1934 Harold Clayton Urey, USA
1935 Frédéric Joliot, France; Irène Joliot-Curie, France
1936 Peter Josephus Wilhelmus Debye, The Netherlands
1937 Walter Norman Haworth, UK; Paul Karrer, Switzerland
1938 Richard Kuhn, Germany
1939 Adolf Friedrich Johann Butenandt, Germany; Leopold Ruzicka, Switzerland
1940 No award
1941 No award
1942 No award
1943 George de Hevesy, Hungary
1944 Otto Hahn, Germany
1945 Artturi Ilmari Virtanen, Finland
1946 James Batcheller Sumner, USA; John Howard Northrop, USA; Wendell Meredith Stanley, USA
1947 Sir Robert Robinson, UK
1948 Arne Wilhelm Kaurin Tiselius, Sweden

1949 William Francis Giauque, USA
1950 Otto Paul Hermann Diels, Germany; Kurt Adler, Germany
1951 Edwin Mattison McMillan, USA; Glenn Theodore Seaborg, USA
1952 Archer John Porter Martin, UK; Richard Laurence, UK
1953 Hermann Staudinger, Germany
1954 Linus Carl Pauling, USA
1955 Vincent du Vigneaud, USA
1956 Cyril Norman Hinshelwood, UK; Nikolai Nikolaevich Semenov, Soviet Union
1957 Alexander R. Todd, UK
1958 Frederick Sanger, UK
1959 Jaroslav Heyrovsky, Czechoslovakia
1960 Willard Frank Libby, USA
1961 Melvin Calvin, USA
1962 Max Ferdinand Perutz, UK; John Cowdery Kendrew, UK
1963 Karl Ziegler, Germany; Giulio Natta, Italy
1964 Dorothy Crowfoot Hodgkin, UK
1965 Robert Burns Woodward, USA
1966 Robert S. Mulliken, USA
1967 Manfred Eigen, Austria; Ronald George Wreyford Norrish, UK; George Porter, UK
1968 Lars Onsager, USA
1969 Derek H. R Barton, UK; Odd Hassel, Norway
1970 Luis F. Leloir, Argentina
1971 Gerhard Herzberg, Canada
1972 Christian B. Anfinsen, USA; Stanford Moore, USA; William H. Stein, USA
1973 Ernst Otto Fischer, Germany; Geoffrey Wilkinson, UK
1974 Paul J. Flory, USA
1975 John Warcup Cornforth, Australia and UK; Vladimir Prelog, Switzerland
1976 William N. Lipscomb, USA
1977 Ilya Prigogine, Belgium
1978 Peter D. Mitchell, UK
1979 Herbert C. Brown, USA; George Witting, Austria
1980 Paul Berg, USA; Walter Gilbert, USA; Frederick Sanger, UK
1981 Kenichi Fukui, Japan; Roald Hoffmann, USA
1982 Aaron Klug, UK
1983 Henry Taube, USA
1984 Robert Bruce Merrifield, USA
1985 Herbert A. Hauptman, USA; Jerome Karle, USA
1986 Dudley R. Herschbach,USA; Yuan T. Lee, USA; John C. Polanyi, Canada
1987 Donald J. Cram, USA; Jean-Maire Lehn, France; Charles J. Pedersen, USA
1988 Johann Deisenhofer, Germany; Robert Hubert, Germany; Hartmut Michel, Germany
1989 Sidney Altman, Canada and USA; Thomas R. Cech, USA
1990 Elias James Corey, USA
1991 Richard E. Ernst, Switzerland
1992 Rudolph A. Marcus, USA
1993 Kary B. Mullis, USA; Michael Smith, Canada
1994 George A. Olah, USA
1995 Paul J. Crutzen, The Netherlands; Mario J. Molina, USA; F. Sherwood Rowland, USA
1996 Robert F. Curl Jr, USA; Sir Harold W. Kroto, UK; Richard E. Smalley, USA
1997 Paul D. Boyer, USA; John E. Walker, UK; Jens C. Skou, Denmark
1998 Walter Kohn,USA; John A. Pople, UK
1999 Ahmed H. Zewail, Egypt and USA
2000 Alan J. Heeger, USA; Alan G. MacDiarmid, USA and New Zealand; Hideki Shirakawa, Japan

2001 William S. Knowles, USA; Ryoji Noyori, Japan; K. Barry Sharpless, USA

Physiology/Medicine

1901 Emil Adolf von Behring, Germany
1902 Ronald Ross, UK
1903 Niels Ryberg Finsen, Denmark
1904 Ivan Petrovich Pavlov, Russia
1905 Robert Kand, Germany
1906 Camillo Golgi, Italy; Santiago Ramón y Cajal, Spain
1907 Charles Louis Alphonse Laveran, France
1908 Ilya Ilyich Metchnikov, Russia; Paul Ehrlich, Germany
1909 Emil Theodor Kander, Switzerland
1910 Albrecht Kossel, Germany
1911 Allvar Gullstrand, Sweden
1912 Alexis Carrel, France
1913 Charles Robert Richert, France
1914 Robert Bárány, Austria
1915 No award
1916 No award
1917 No award
1918 No award
1919 Jules Bordet, Belgium
1920 Schack August Steenberg Krogh, Denmark
1921 No award
1922 Archibald Vivian Hill, UK; Otto Frits Eyerhof, Germany
1923 Fredrick Grant Banting, Canada; John James Richard Macleod, Canada
1924 Willem Einthoven, The Netherlands
1925 No award
1926 Johannes Andreas Grib Fibiger, Denmark
1927 Julius Wagner-Jauregg, Austria
1928 Charles Jules Henri Nicolle, France
1929 Christiaan Eijkman, The Netherlands; Sir Frederick Hopkins, UK
1930 Karl Landsteiner, Austria
1931 Otto Heinrich Warburg, Germany
1932 Sir Charles Scott Sherrington, UK; Edgar Douglas Adrian, UK
1933 Thomas Hunt Morgan, USA
1934 George Hoyt Whipple, USA; George Richard Minot, USA; William Parry Murphy, USA
1935 Hans Spemann, Germany
1936 Sir Henry Hallet Dale, UK; Otto Loewi, Austria
1937 Albert von Szent-Györgyl Nagyrapolt, Hungary
1938 Corneille Kean Francois Heymans, Belgium
1939 Gerhard Domagk, Germany
1940 No award
1941 No award
1942 No award
1943 Henrick Carl Peter Dam, Denmark; Edward Adelbert Doisy, USA
1944 Joseph Erlanger, USA; Herbert Spencer Gasser, USA
1945 Sir Alexander Fleming, UK; Ernst Boris Chain, UK; Sir Howard Walter Florey, Australia
1946 Hermann Joseph Muller, USA
1947 Cark Ferdinand Cori, USA; Gerty Theresa Cori, née Radnitz, USA; Bernardo Alberto Houssay, Argentina
1948 Paul Hermann Müller, Switzerland
1949 Walter Hess, Switzerland; Antonio Caetano de Abreu Freire Egas Moniz, Portugal
1950 Edward Calvin Kendall, USA; Tadeus Reichstein, Switzerland; Philip Showalter Hench, USA
1951 Max Theiler, South Africa
1952 Selman Abraham Waksman, USA
1953 Hans Adolf Krebs, UK; Fritz Albert Lipmann, USA

1954 Johan Franklind Enders, USA; Thomas Huckle Weller, USA; Frederick Chapman Robbins, USA
1955 Axel Hugo Theodor Theorell, Sweden
1956 André Frédéric Cournand, USA; Werner Forssman, Germany; Dickinson W. Richards, USA
1957 Daniel Bovet, Italy
1958 George Wells Beadie, USA; Edward Tatum, USA; Joshua Lederberg, USA
1959 Severo Ochoa, USA; Arthur Kornberg, USA
1960 Sir Frank Macfarlane Burnet, Australia; Peter Brian Medawar, UK
1961 Georg von Békésy, USA
1962 Francis Harry Compton Crick, UK; James Dewey Watson, USA; Maurice Hugh Frederick Wilkins, USA and New Zealand
1963 Sir Jon Carew Eccles, Australia; Alan Lloyd Hodgkin, UK; Andrew Fielding Huxley, UK
1964 Konrad Bland, USA; Feodor Lynen, Germany
1965 Francoise Jacob, France; André Lwoff, France; Jacques Monod, France
1966 Peyton Rous, USA; Charles Brenton Huggins, USA
1967 Ragnar Granit, Sweden; Haldan Keffer Hartline, USA; George Wald, USA
1968 Robert W. Holley, USA; Har Gobind Khorana, USA; Marshall W. Nirenberg, USA
1969 Max Delbrück, USA; Alfred D. Hershey, USA; Salvador E.Luria, USA
1970 Sir Bernard Katz, UK; Ulf von Euler, Sweden; Julius Axelrod, USA
1971 Earl W Sutherland, Jr, USA
1972 Gerald M Edelman, USA; Rodney R. Porter, UK
1973 Karl von Frisch, Germany; Konrad Lorenz, Austria; Nikolaas Tinbergen, UK
1974 Albert Claude, Belgium; Christian de Duve, Belgium; George E. Palade, USA
1975 David Baltimore, USA; Renato Dulbecco, USA; Howard Martin Temin, USA
1976 Baruch S. Blumberg, USA; D. Carleton Gajdusek, USA
1977 Roger Guillemin, USA; Andrew V. Schally, USA; Rosalyn Yalow, USA
1978 Werner Aber, Switzerland; Daniel Nathans, USA; Hamilton O. Smith, USA
1979 Allan M. Cormarck, USA; Godfrey N. Hounsfield, UK
1980 Baruj Benacerraf, USA; Jean Dausset, France; George D. Snell, USA
1981 Roger W. S. Perry, USA; David H. Hubel, USA; Torsten N. Wiesel, Sweden
1982 Sune K. Bergström, Sweden; Bengt I. Samuelsson, Sweden; John R. Vane, UK
1983 Barbara McClintock, USA
1984 Niels K. Jerne, Denmark; Georges J. F. Köhler, Germany; César Milstein, Argentina and UK
1985 Michael S. Brown, USA; Joseph L. Goldstein, USA
1986 Stanley Cohen, USA; Rita Levi. Monalcini, Italy and USA
1987 Susumu Tonegawa, Japan
1988 Sir James W. Black, UK; Gertrude B. Elion, USA; George H. Hitchings, USA
1989 J. Michael Bishop, USA; Harold E. Varmus, USA
1990 Joseph E. Murray, USA; E. Donnall Thomas, USA
1991 Erwin Neher, Germany; Bert Sakmann, Germany

1992	Edmond H. Fischer, Switzerland and USA; Edwin G. Krebs, USA
1993	Richard J. Roberts, UK; Phillip A. Sharp, USA
1994	Alfred G. Gilman, USA; Martin Rodbell, USA
1995	Edward B. Lewis, USA; Christiane Nüsslein-Volhard, Germany; Eric F. Wieschaus, USA
1996	Peter C. Doherty, Australia; Rolf M. Zinkernagel, Switzerland
1997	Stanley B. Prusiner, USA
1998	Robert F. Furchgott, USA; Louis J. Ignarro, USA; Ferid Murad, USA
1999	Günter Blobel, USA
2000	Arvid Carlsson, Sweden; Paul Greengard, USA; Eric R. Kandel, USA
2001	Leland H. Hartwell, USA; R. Timothy Hunt, UK; Sir Paul M. Nurse, UK

Literature

1901	Sully Prudhomme, France
1902	Theodor Mommsen, Germany
1903	Bjönstjerne Björnson, Norway
1904	Frédéric Mistral, France; José Echegaray, Spain
1905	Henryk Sienkiewicz, Poland
1906	Giosué Carducci, Italy
1907	Rudyard Kipling, UK
1908	Rudolf Eucken, Germany
1909	Selma Lagerlöf, Sweden
1910	Paul Heyse, Germany
1911	Maurice Maeterlinck, Belgium
1912	Gerhart Hauptmann, Germany
1913	Rabindranath Tagore, India
1914	No award
1915	Romain Rolland, France
1916	Verner von Heidenstam, Sweden
1917	Karl Gjellerup, Denmark; Henrik Pontopiddan, Denmark
1918	No award
1919	Carl Spitteler, Switzerland
1920	Knut Hamsun, Norway
1921	Anatole France, France
1922	Jacinto Benavente, Spain
1923	William Butler Yeats, Ireland
1924	Wladyslaw Reymint, Poland
1925	George Bernard Shaw, UK
1926	Grazia Deledda, Italy
1927	Henri Bergson, France
1928	Sigrid Undset, Norway
1929	Thomas Mann, Germany
1930	Sinclair Lewis, USA
1931	Erik Axel Karlfeldt, Sweden
1932	John Galsworthy, UK
1933	Ivan Bunin, France
1934	Luigi Pirandello, Italy
1935	No award
1936	Eugene O'Neill, USA
1937	Roger Martin du Gard, France
1938	Pearl Buck, USA
1939	Frans Eemil Sillanpää, Finland
1940	No award
1941	No award
1942	No award
1943	No award
1944	Johannes v. Jensen, Denmark
1945	Gabriela Mistral, Chile
1946	Herman Hesse, Switzerland
1947	André Gide, France
1948	Thomas Stearns Eliot, UK
1949	William Faulkner, USA
1950	Bertrand Russell, UK
1951	Pär Lagerkvist, Sweden
1952	Francois Mauriac, France
1953	Winston Churchill, UK
1954	Ernest Hemingway, USA
1955	Halidór Kiljan Laxness, Iceland
1956	Juan Ramón Jiménez, Spain
1957	Albert Camus, France
1958	Boris Pasternak, Soviet Union
1959	Salvatore Quasimodo, Italy
1960	Saint-John Perse, France
1961	Ivo Andric, Jugoslavia
1962	John Steinbeck, USA
1963	Giorgos Seferis, Greece
1964	Jean-Paul Sartre, France

1965 Mikhail Sholokhov, Soviet Union
1966 Samuel Agnon, Israel; Nelly Sachs, Sweden
1967 Miguel Angel Asturias, Guatemala
1968 Yasunari Kawabata, Japan
1969 Samuel Beckett, Ireland
1970 Alexandr Solzhenitsyn, Soviet Union
1971 Pablo Neruda, Chile
1972 Heinrich Böll, Germany
1973 Patrick White, Australia
1974 Eyvind Johnson, Sweden; Harry Martinson, Sweden
1975 Eugenio Montale, Italy
1976 Saul Bellow, USA
1977 Vicente Aleixandre, Spain
1978 Isaac Bashevis Singer, USA
1979 Odysseus Elytis, Greece
1980 Czeslaw Milosz, Poland and USA
1981 Elias Canetti, UK
1982 Gabriel Garcia Márquez, Colombia
1983 William Golding, UK
1984 Jaroslav Seifert, Czechoslovakia
1985 Claude Simon, France
1986 Wole Soyinka, Nigeria
1987 Joseph Brodsky, USA
1988 Naguib Mahfouz, Egypt
1989 Camilo José Cela, Spain
1990 Octavio Paz, Mexico
1991 Nadine Gordimer, South Africa
1992 Derek Walcott, St Lucia
1993 Toni Morrison, USA
1994 Kenzaburo Oe, Japan
1995 Seamus Heaney, Ireland
1996 Wislawa Szymborska, Poland
1997 Dario Fo, Italy
1998 José Saramango, Portugal
1999 Günter Grass, Germany
2000 Gao Xingjian, France
2001 V.S. Naipaul, UK

Economic Sciences

1969 Ragnar Frisch, Norway; Jan Tinbergen, The Netherlands
1970 Paul A. Samuelson, USA
1971 Simon Kuznets, USA
1972 John R. Hicks, UK; Kenneth J. Arrow, USA
1973 Wassily Leontief, USA
1974 Gunnar Myrdal, Sweden; Friedrich August von Hayek, UK
1975 Leonid Vitaliyevich Kantarovich, Soviet Union; Tjalling C. Koopmans, USA
1976 Milton Friedman, USA
1977 Bertil Ohlin, Sweden; James E. Meade, UK
1978 Herbert A. Simon, USA
1979 Theodore W. Schultz, USA; Sir Arthur Lewis, UK
1980 Lawrence R. Klein, USA
1981 James Tobin, USA
1982 George J. Stigler, USA
1983 Gerard Debreu, USA
1984 Richard Stone, UK
1985 Franco Modigliani, USA
1986 James M. Buchanan Jr, USA
1987 Robert M. Solow, USA
1988 Maurice Allais, France
1989 Trygve Haavelmo, Norway
1990 Harry M. Markowitz, USA; Merton H. Miller, USA; William F. Sharpe, USA
1991 Ronald H. Coase, UK
1992 Gary S. Becker, USA
1993 Robert W. Fogel, USA; Douglass C. North, USA
1994 Johan C. Harsanyi, USA; John F. Nash Jr, USA; Reinhard Selter, Germany
1995 Robert E. Lucas Jr, USA
1996 James A. Mirrlees, UK; William Vickrey, USA
1997 Robert C. Merton, USA; Myron S. Scholes, USA